Target
Get back on track

GRADE **3**

Edexcel GCSE (9–1)
Mathematics
Number and Statistics

Diane Oliver

Pearson

Published by Pearson Education Limited, 80 Strand, London, WC2R 0RL.

www.pearsonschoolsandfecolleges.co.uk

Text © Pearson Education Limited 2017
Typeset by Tech-Set Ltd, Gateshead
Original illustrations © Pearson Education Ltd 2017

The right of Diane Oliver to be identified as author of this work has been asserted by her in accordance with the
Copyright, Designs and Patents Act 1988.

First published 2017

19 18 17
10 9 8 7 6 5 4 3 2 1

British Library Cataloguing in Publication Data
A catalogue record for this book is available from the British Library

ISBN 978 0 435 18332 5

Printed in Italy by Lego S.p.A

Helping you to formulate grade predictions, apply interventions and track progress.

Any reference to indicative grades in the Pearson Target Workbooks and Pearson Progression Services is not to be used
as an accurate indicator of how a student will be awarded a grade for their GCSE exams.

You have told us that mapping the Steps from the Pearson Progression Maps to indicative grades will make it simpler
for you to accumulate the evidence to formulate your own grade predictions, apply any interventions and track student
progress.

We're really excited about this work and its potential for helping teachers and students. It is, however, important to
understand that this mapping is for guidance only to support teachers' own predictions of progress and is not an
accurate predictor of grades.

Our Pearson Progression Scale is criterion referenced. If a student can perform a task or demonstrate a skill, we say
they are working at a certain Step according to the criteria. Teachers can mark assessments and issue results with
reference to these criteria which do not depend on the wider cohort in any given year. For GCSE exams however, all
Awarding Organisations set the grade boundaries with reference to the strength of the cohort in any given year. For
more information about how this works please visit: https://qualifications.pearson.com/en/support/support-topics/results-
certification/understanding-marks-and-grades.html/Teacher

Each practice question features a Step icon which denotes the level of challenge aligned to the
Pearson Progression Map and Scale.

To find out more about the Progression Scale for Maths and to see how it relates to indicative GCSE 9–1
grades go to www.pearsonschools.co.uk/ProgressionServices

Contents

Useful formulae   iv

Glossary   v

Unit 1 Multiplying

   Get started   1

1 Multiplying by 10, 100, 1000, 0.1 and 0.01   2

2 Multiplying decimals by a single-digit
   number   3

3 Multiplying negative integers   4

   Practise the methods   5

   Problem-solve!   6

Unit 2 Dividing

   Get started   7

1 Dividing 3-digit numbers by a single-digit
   number   8

2 Dividing decimals by a single-digit number   9

3 Dividing by 0.1 and 0.01   10

4 Dividing negative integers   11

   Practise the methods   12

   Problem-solve!   13

Unit 3 Fractions

   Get started   14

1 Mixed numbers and improper fractions   15

2 Adding and subtracting fractions and mixed
   numbers   16

3 Multiplying a fraction by an integer   17

4 Dividing an integer by a fraction   18

   Practise the methods   19

   Problem-solve!   20

Unit 4 Fractions, decimals and percentages

   Get started   21

1 Converting between decimals and fractions   22

2 Converting a fraction to a decimal   23

3 Writing one number as a percentage of
   another   24

4 Ordering and comparing fractions, decimals
   and percentages   25

   Practise the methods   26

   Problem-solve!   27

Unit 5 Probability

   Get started   28

1 The probability scale   29

2 Mutually exclusive outcomes for one event   30

3 Estimating successes   31

4 Mutually exclusive outcomes for two events
   and frequency trees   32

   Practise the methods   34

   Problem-solve!   35

Unit 6 Ratio and proportion

   Get started   36

1 Simplifying and using ratio   37

2 Using proportion   38

3 Inverse proportion   39

   Practise the methods   40

   Problem-solve!   41

Unit 7 Averages and range

   Get started   42

1 Averages and range   43

2 Comparing two distributions   45

3 Data in tables   47

   Practise the methods   49

   Problem-solve!   51

Unit 8 Data collection

   Get started   52

1 Data collection sheets   53

2 Bias   55

   Practise the methods   57

   Problem-solve!   58

Unit 9 Tables, charts and graphs

   Get started   59

1 Dual bar charts and line graphs   60

2 Interpreting pie charts   62

3 Drawing pie charts   63

4 Misleading graphs   64

   Practise the methods   65

   Problem-solve!   66

Answers   68

Useful formulae

Unit 1 Multiplying

Rules for multiplying integers:

positive × positive = positive

positive × negative = negative

negative × negative = positive

negative × positive = negative

Unit 2 Dividing

Rules for dividing integers:

positive ÷ positive = positive

positive ÷ negative = negative

negative ÷ negative = positive

negative ÷ positive = negative

Unit 5 Probability

Probability of an outcome $= \dfrac{\text{number of ways outome can happen}}{\text{total number of possible outcomes}}$

Sum of probabilities of mutually exclusive outcomes = 1

Estimate of number of successes = probability of success × number of trials

Unit 7 Averages and range

Mean $= \dfrac{\text{sum of data values}}{\text{number of data values}}$

Range = highest data value − lowest data value

Glossary

1 Multiplying

Decimal: a number written using the base ten number system, usually including numbers after the decimal point.

Integer: positive or negative number that is not a fraction or decimal with numbers after the decimal point.

Place value: digits have a value that depends on their position (place) in the number. So in 500, the digit 5 has the value 500, but in 50, the digit 5 has the value 50.

3 Fractions

Denominator: the bottom number in a fraction.

Equivalent fraction: fractions that look different but have the same value. For example $\frac{2}{6}$ and $\frac{1}{3}$.

Improper fraction: a fraction where the numerator is greater than the denominator. For example, $\frac{5}{3}$

Mixed number: a number that includes a whole number part and a fraction part. For example, $1\frac{1}{3}$

Numerator: the top number in a fraction.

Proper fraction: a fraction where the numerator is less than the denominator. For example, $\frac{2}{3}$

Reciprocal: the reciprocal of a number is 1 divided by the number. The reciprocal of a fraction turns it upside down. For example, the reciprocal of $\frac{2}{3}$ is $\frac{3}{2}$ and the reciprocal of 3 ($= \frac{3}{1}$) is $\frac{1}{3}$

4 Fractions, decimals and percentages

Descending: going down in value.

Percentage: a fraction expressed as 'out of 100'.

5 Probability

Biased: gives more of a particular outcome than expected. For example, a biased die could give more '4s' than you might expect.

Frequency tree: a diagram that shows the number of options for different outcomes.

Mutually exclusive: mutually exclusive events cannot happen at the same time. For example, when flipping a coin, getting a head or a tail from one coin are mutually exclusive events because you cannot get a head and a tail at the same time from one coin.

Probability scale: a scale showing probabilities. It is labelled from 0 to 1 and is used to position and compare probabilities.

Two-way table: a way of listing outcomes for two or more events.

6 Ratio and proportion

Highest common factor: if you find all the factors of two numbers, the highest number that is a factor of both numbers is the highest common factor (HCF).

Inverse proportion: where one value increases at the same rate as another decreases.

Simplest form: a form where the numbers in the ratio cannot be divided by any further common factors. To find the simplest form of a ratio, you divide the numbers by their highest common factor.

Unitary method: a method where first you work out the answer for 1 item, then use this value to work out the answer for multiple items.

7 Averages and range

Mean: the value obtained when all the values in a data set are added together and then the total is divided by the number of items. For example, in the data set 1, 2, 2, 3, 4, 5, 6, the mean is $\frac{1 + 2 + 2 + 3 + 4 + 5 + 6}{7} = 3.2857\ldots$

Median: the middle value of the data set when the set is written in order. For example, in the data set 1, 2, 2, 3, 4, 5, 6, the median is 3

Modal class: when data is arranged in a frequency table, the modal class is the class that has the highest frequency.

Mode: the most common value in the data set. For example, in the data set 1, 2, 2, 3, 4, 5, 6, the mode is 2

Range: the difference between the smallest and largest values in a data set. The range shows how spread out a data set is. For example, in the data set 1, 2, 2, 3, 4, 5, 6, the range is $6 - 1 = 5$

8 Data collection

Bias: anything that may influence the outcome of a survey.

Leading question: a question on a survey that tries to influence your answer.

Class interval: the size of an interval defining a group (class) on a data collection sheet. For example, if the classes on a data collection sheet about number of texts received were 1–5, 6–10, 11–15, etc., the class interval is 5

Continuous data: data that is measured and can have any value. Examples are time, length, and weight.

Data collection sheet: a table used to collect data.

Discrete data: data that contains particular values. For example, the number of texts you receive in a week is discrete data.

Fair: something that isn't biased.

Grouped data: data that has been put into groups. For example, you might group number of texts received in a week into groups 1–5, 6–10, 11–15, etc.

Hypothesis: a statement that you are trying to prove or disprove by collecting data.

Questionnaire: a method of collecting data by asking questions.

Sample: a sample represents the population being studied.

Population: the whole group that you are interested in.

9 Tables, charts and graphs

Dual bar chart: the height of a bar on a bar chart represents frequency. A dual bar chart compares two sets of data, for example, boys with girls. The bars for the same categories are drawn side by side.

Key: a legend at the side of a statistical diagram showing what is represented on the diagram.

Line graph: a way of representing continuous data by joining points.

Pie chart: a way of representing data using sectors of a circle. The size of the sectors relates to the frequency of the data being represented by each sector.

① Multiplying

This unit will help you to multiply decimals and integers.

① Write down the value of the 2 in each number.

a 253   **b** 9721   **c** 432   **d** 32 486

[handwritten: H T U] 200 ✓   *[handwritten: Th H T U]* 20 units ✓   2 units ✓   2 thousand

② Write down the value of the 2 in each number.

a 2.7   **b** 6.2   **c** 7.512   **d** 4.829

2 units ✓   2 tenths ✓   2 thousandths ✓   2 hundredths ✓

③ **Number sense**

Work out

a 3 × 100   **b** 45 × 10   **c** 2 × 1000   **d** 67 × 100

300 ✓   450 ✓   2000 ✓   6700 ✓

Key points

Multiplication can make a number smaller, e.g. 40 × 0.1 = 4

Integers are positive and negative numbers that are not decimals or fractions, e.g. −26, −5, 1 and 8.

These **skills boosts** will help you to multiply positive and negative decimals and integers.

1 Multiplying by 10, 100, 1000, 0.1 and 0.01

2 Multiplying decimals by a single-digit number

3 Multiplying negative integers

You might have already done some work on multiplication. Before starting the first skills boost, rate your confidence using each concept.

① Work out 583 × 0.01

5.83 ×
0
5.83 ✓

② Work out 4.72 × 6

28.32 ✓

③ Work out −7 × 9

−63 ✓

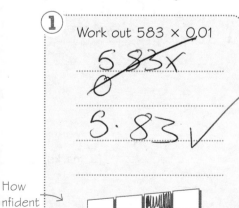

How confident are you?

1 Multiplying by 10, 100, 1000, 0.1 and 0.01

Multiplying by 0.1 is the same as dividing by 10, so it makes the number you are multiplying 10 times smaller. Multiplying by 0.01 is the same as dividing by 100, so it makes the number you are multiplying 100 times smaller.

Guided practice

Work out　**a** 14.52 × 1000　　**b** 9617 × 0.01

a Write the number in a place-value table.
Each time you multiply by 10, move the digits one place to the left in the table.
Write 0 in empty columns to the left of the decimal point.

	TTh	Th	H	T	U	.	$\frac{1}{10}$	$\frac{1}{100}$
				1	4	.	5	2
× 10 =			1	4	5	.	2	
× 100 =		1	4	5	2			
× 1000 =	1	4	5	2	0			

Multiplying by 10 makes each digit 10 times bigger.
Multiplying by 100 makes each digit 100 times bigger.
Multiplying by 1000 makes each digit 1000 times bigger.

14.52 × 1000 = 14520 ✓

b Write the number in a place-value table.
Each time you divide by 10, move the digits one place to the right in the table.

	Th	H	T	U	.	$\frac{1}{10}$	$\frac{1}{100}$
	9	6	1	7			
× 0.1 =		9	6	1	.	7	
× 0.01 =			9	6	.	1	7

Multiplying by 0.1 makes each digit 10 times smaller.
Multiplying by 0.01 makes each digit 100 times smaller.

9617 × 0.01 = 96.17

1 Work out
　a 3.7 × 10 = _37_ ✓　　**b** 1.8 × 10 = _18_ ✓　　**c** 56.4 × 100 = _5640_ ✓
　d 3.85 × 10 = _385_ ✓　**e** 9.37 × 100 = _9370_ ✓　**f** 7.6 × 1000 = _7600_ ✓

2 Work out
　a 0.2 × 10 = _2_ ✓　　**b** 0.45 × 100 = _450_ ✓　　**c** 0.6 × 1000 = _600_ ✓

3 Work out　**Hint** 30.0 is the same as 30.
　a 300 × 0.1= _30_ ✓　　**b** 42 × 0.1= _4.02_ ✓　　**c** 735 × 0.1= _73.5_ ✓
　d 9.1 × 0.1= _0.91_ ✓　**e** 400 × 0.01= _4_ ✓　　**f** 296 × 0.01= _2.96_ ✓
　g 52 × 0.01= _0.52_ ✓　**h** 472.1 × 0.01= _4.72_ ✓　**i** 3.8 × 0.01= _0.038_ ✓

Exam-style question

4 Chris is going on holiday to Taiwan.
He wants to change £100 into New Taiwan dollars.
The exchange rate is £1 = 38.64 dollars.
How many New Taiwan dollars should he get?

3864

(2 marks)

Reflect　Explain how to multiply by 0.01.

② Multiplying decimals by a single-digit number

For any multiplication, always use an estimate to check your answer.

Guided practice

Work out 5.8 × 6

Ignore the decimal point and work out 58 × 6.

$$\begin{array}{r} 5\;8 \\ \times\;\;6 \\ \hline 34\;8 \end{array}$$

58 × 6 = 348

5.8 × 6 = 34.8

Check: 5.8 ≈ 6, 6 × 6 = 36 ✓

Handwritten:
$$\begin{array}{r} 64\times \\ 7 \\ \hline 448 \end{array}$$

$$\begin{array}{r} 51\times \\ 9 \\ \hline 459 \end{array}$$

$$\begin{array}{r} 59\times \\ 7 \\ \hline 413 \end{array}$$

58 ÷ 10 = 5.8, so divide your answer by 10 to get the final answer.

The estimate is 36. Is your answer close to 36?

① Work out

a 4.3 × 8
34.4 ✓

b 7.2 × 5 *(handwritten 360 / 360)*
36 ✓

Hint 9 × 5.1 is the same as 5.1 × 9.

c 9 × 5.1   5.1 × 9
45.9 ✓

d 7 × 6.4   44.8 44.8 ✓

e 5.9 × 7
41.3 ✓

f 6 × 9.2
55.2 ✓

Handwritten:
$$\begin{array}{r} 92\times \\ 6 \\ \hline 552 \end{array}$$

② Work out

Hint Ignore the decimal point and work out 327 × 5.
327 ÷ 100 = 3.27, so divide your answer by 100 to get the final answer.

a 3.27 × 5
16.35 ✓

b 4.81 × 7
33.67 ✓

c 7.38 × 6
44.28 ✓

d 9 × 5.16
46.44 ✓

e 8.07 × 8
64.56 ✓

f 4 × 7.09
28.36 ✓

③ Work out

a 9.5 × 7
66.5 ✓

b 8.24 × 6
49.44 ✓

c 8 × 1.67
13.36 ✓

d 7.3 × 4
29.2 ✓

e 2 × 9.48
18.96 ✓

f 6 × 5.7
34.2 ✓

Exam-style question

④ Work out

a £4.85 × 6
£29.1 ✓ *(2 marks)*

Handwritten:
$$\begin{array}{r} 485\times \\ 6 \end{array}$$

b £8.30 × 5
£41.5 *(2 marks)*

Handwritten:
$$\begin{array}{r} 830\times \\ 5 \\ \hline 4150 \end{array}$$

Reflect

Look at your methods for answering Q1b and Q1c. Are they different? Is one question more difficult than the other?

3 Multiplying negative integers

The rules for multiplying integers are:
- positive × positive = positive
- positive × negative = negative
- negative × positive = negative
- negative × negative = positive

Guided practice

Work out 3 × −4

First work out the calculation without the signs.

3 × 4 = ___12___

Use the rules for multiplying negative numbers.

positive × negative = ___−12___

Write the answer with the correct sign.

3 × −4 = −12

3 × −4 means 3 lots of −4,
so −4 + −4 + −4 or:

| −1 | −1 | −1 | −1 |

+

| −1 | −1 | −1 | −1 |

+

| −1 | −1 | −1 | −1 |

① **a** Complete the multiplication grid.

b Explain what you notice about the numbers in the pink boxes.
___All Numbers are___

c Explain what you notice about the numbers in the blue boxes.
___they are negative___

d Explain what you notice about the numbers in the white boxes.
___They are positive___

×	−4	−3	−2	−1	0	1	2	3	4
4	0	1	2	3	0	4	8	12	16
3	−1	0	1	2	0	3	6	9	12
2	−2	−1	0	1	0	2	4	6	8
1	−3	−2	−1	0	0	1	2	3	4
0	−4	−3	−2	−1	0	1	2	3	4
−1	−5	−4	−3	−2	0	−1	−2	−3	−4
−2	−6	−5	−4	−3	0	−2	−4	−6	−8
−3	−7	−6	−5	−4	0	−3	−6	−9	−12
−4	−8	−7	−6	−5	0	−4	−8	−12	−16

② Work out

a 5 × −3 = ___2___ ___−12___

b −2 × −11 = ___−13___

c −9 × 4 = ___−36___

d 6 × −7 = ___−42___

e 11 × 9 = ___99___

f −3 × 6 = ___−18___

g 8 × −2 = ___6___

h −8 × −12 = ___−20___

③ Work out   **Hint** Multiply the decimal and the integer and then insert the sign.

a 37 × −0.1 = ___36.09___

b −612 × 0.01 = _____

c −8357 × −0.01 = _____

d 6.4 × −3 = _____

e −9 × 7.1 = _____

f −4.5 × 9 = _____

g −6.7 × −5 = _____

h 4 × −8.9 = _____

i −3 × −1.8 = _____

Exam-style question

④ Work out

a 3 × −9   _____ (1 mark)   **b** −12 × −7   _____ (1 mark)

Reflect Without looking at this page, write the rules for multiplying positive and negative numbers.

Practise the methods

Answer this question to check where to start.

Check up

Tick the correct answer for 3.85 × 1000

 A 385 000 ○   **B** 3.850 ○   **C** 3850 ✓ **D** 385 ○

If you ticked C go to Q4. ·

If you ticked A, B or D go to Q1 for more practice.

1 Multiply each number by 10.

a 3   30

Hint

	H	T	U
			3
× 10		3	0

b 1.8   18

Hint

	H	T	U	.	1/10
			1	.	8
× 10		1			

c 12.62   126.02

Hint

	H	T	U	.	1/10	1/100
		1	2	.	6	2
× 10						

2 Multiply each number by 100.

a 5   500

b 3.82   382

c 2.1   210

Hint

	H	T	U	.	1/10
			2	.	1
× 100	2				

3 Multiply each number by 1000.

a 8   8000

b 12.35   12350

c 5.2   5200

4 Work out

a 17 × 0.1   1.7

b 3.2 × −0.1   −0.32

c −56 × −0.01   0.56

5 Work out

a 6.7 × 2   13.4

b 4.1 × 9   36.9

c −6 × 0.58   −3.48

Exam-style question

6 Work out

a 5.7 × 1000   5700 (1 mark)

b 4.9 × 7   34.3 (2 marks)

Problem-solve!

Exam-style questions

1. Emily is going on holiday to Thailand. She needs to change some money into Thai baht.
   Emily wants to change up to £300 into baht notes.
   She wants as many 100 baht notes as possible.
   The exchange rate is £1 = 43.66 Thai baht.
   How many 100 baht notes should she get?

   130.98

   130 N
   ~~130~~

   (5 marks)

2. Grace is going to buy exactly 10 ink cartridges.

   Find the <u>difference</u> in cost between the cheapest way and the most expensive way to buy the 10 ink cartridges.

   £13.20×10=£132
   £31.71×3+£13.20=£108.33
   £132−£108.33=£23.67

   £13.20 per cartridge

   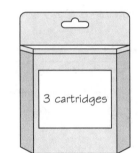

   3 cartridges

   £31.71 for a pack of 3 cartridges

   (5 marks)

3. The table shows the temperatures in some cities at midnight on a night in January.

City	Edinburgh	Newcastle	Norwich	Bangor	Truro	Brighton	London
Temperature (°C)	−5	−6	−3	2	3	−2	6

   Which city was three times as cold as Brighton at midnight?

   Newcastle (1 mark)

4. The table gives information about the costs of posting parcels.

Maximum weight of parcel	Cost
2 kg	£4.89
5 kg	£13.75
10 kg	£20.25
20 kg	£28.55

   Ewan needs to post
   - 3 parcels with a weight of 1.2 kg each
   - 4 parcels with a weight of 8 kg each.

   He has £95 to spend on posting the parcels.
   Can he post all the parcels?

   Yes (3 marks)

5. Mia and Niall work at the same restaurant.
   You can use this formula to work out the total amount of money each person gets.

   Total amount = £5.25 × number of hours worked + tips

	Number of hours worked	Tips
Mia	8	£12
Niall	7	£15

   The table shows the number of hours Mia and Niall each worked last Saturday and the tips they got.

   Who got the higher total amount of money?
   You must show clearly how you got your answer.

   £5.25×8+£12=£54
   £5.25×7+£15=£51.75

   Mia (4 marks)

Now that you have completed this unit, how confident do you feel?

1. Multiplying by 10, 100, 1000, 0.1 and 0.01

2. Multiplying decimals by a single-digit number

3. Multiplying negative integers

② Dividing

This unit will help you to divide decimals and integers.

AO1 Fluency check

① Work out

a 24 ÷ 6 = *4*   **b** 45 ÷ 5 = *9*   **c** 27 ÷ 3 = *9*

d 32 ÷ 4 = *8*   **e** 56 ÷ 8 = *7*   **f** 42 ÷ 7 = *6*

② Use a written method to work out

a 46 ÷ 2   **b** 75 ÷ 5   **c** 72 ÷ 3   **d** 92 ÷ 4

23   *15*   *24*   *23*

③ Number sense

Work out

a 5 ÷ 10   **b** 47 ÷ 10   **c** 12.7 ÷ 10

0.5   *4.7*   *1.27*

d 354.2 ÷ 10   **e** 368.9 ÷ 100   **f** 23 ÷ 100

35.42   *3.689*   *0.23*

g 9675.8 ÷ 100   **h** 35.6 ÷ 1000   **i** 4007.1 ÷ 1000

96.758   *0.0356*   *4.0071*

Key points

Division can make a number bigger, e.g. 4 ÷ 0.1 = 40

If the answer to a division calculation has a remainder, it can be written as a decimal.

These **skills boosts** will help you to divide positive and negative decimals and integers.

| 1 Dividing 3-digit numbers by a single-digit number | 2 Dividing decimals by a single-digit number | 3 Dividing by 0.1 and 0.01 | 4 Dividing negative integers |

You might have already done some work on division. Before starting the first skills boost, rate your confidence using each concept.

①	②	③	④
Work out 425 ÷ 8	Work out 5.03 ÷ 4	Work out 7.1 ÷ 0.01	Work out −6.51 ÷ 3
5 30125	*1.2575*	*710*	*−2.17*

How confident are you?

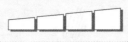

1 Dividing 3-digit numbers by a single-digit number

Guided practice

Work out 462 ÷ 8

Write the division.

Write a decimal point and zeros after 462.

Put a decimal point in the answer line. Line it up with the decimal point underneath.

```
   0 5 7.7
8)4⁴6⁶2.⁶0⁵0
```

462 ÷ 8 = 57.75

8 doesn't go into 4. How many 8s are in 46?
8 goes into 46 five times, with 46 − 40 = 6 left over.
8 goes into 62 seven times, with 62 − 56 = 6 left over.
How many 8s are in 60?

(1) Work out

a 434 ÷ 7

62

b 632 ÷ 4

158

c 657 ÷ 9

73

d 540 ÷ 5

108

e 777 ÷ 3

259

f 564 ÷ 6

94

(2) Work out each division. Give each answer as a decimal.

a 631 ÷ 4

151·151

b 853 ÷ 5

170.6

c 562 ÷ 8

70·25

d 942 ÷ 4

235.5

e 887 ÷ 4

221.75

f 365 ÷ 8

45·625

(3) The entrance fee to a theme park for 4 adults costs £173. **Hint** 'Per adult' means 'for each adult'.
How much does it cost per adult?

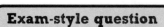 *173÷4*   *173×4 = £692*

Exam-style question

(4) Work out 108 ÷ 8

13·5   (2 marks)

Reflect

Look at your answer to the calculation in Q4. How could you simplify the calculation before working out the division?

2 Dividing decimals by a single-digit number

You divide decimal numbers in the same way as you divide whole numbers. Make sure you line up the decimal points.

Guided practice

Work out 82.75 ÷ 5

Use the same method as for dividing whole numbers.

$$\begin{array}{r} 1\ 6\ .\ 5\ 5 \\ 5\overline{)8\ ^3 2\ .\ ^2 7\ ^2 5} \end{array}$$

Make sure you line up the decimal points.

82.75 ÷ 5 = 16.55

1. Work out

   a 9.8 ÷ 2

   4.9

   b 15.9 ÷ 3

   5.3

   c 16.17 ÷ 7

   2.31

   d 410.94 ÷ 6

   68.49

   e 28.53 ÷ 9

   3.17

   f 6.51 ÷ 3

   2.17

2. Work out **Hint** Write extra zeros if you need more decimal places.

   a 7.3 ÷ 5

   1.46

   b 6.2 ÷ 4

   1.55

   c 5.1 ÷ 2

   2.55

   d 9.3 ÷ 8

   1.1625

   e 7.23 ÷ 4

   1.8075

   f 13.55 ÷ 8

   1.69375

3. Six friends share a bill totalling £74.22.
   Work out how much each person pays.

   £12.37

4. A piece of wood is 5.5 m long.
   Alex cuts the wood into 4 equal pieces.
   How long is each piece?

   £1.375

Exam-style question

5. One day Jane earned £62.80.
   She worked for 8 hours.
   Work out Jane's hourly rate of pay.

   7.85 (2 marks)

Reflect

How can you tell whether you need extra zeros for more decimal places?

3 Dividing by 0.1 and 0.01

Dividing by 0.1 is the same as multiplying by 10, so it makes the number you are dividing 10 times bigger.

Similarly, dividing by 0.01 is the same as multiplying by 100, so it makes the number you are dividing 100 times bigger.

Guided practice

Work out 3.6 ÷ 0.01

Write the number in a place-value table.
When you divide by 0.1, move the digits one place to the left in the table.
When you divide by 0.01, move the digits two places to the left.

H	T	U	.	$\frac{1}{10}$
		3	.	6
× 10	3	6		
× 100	3			

3.6 ÷ 0.01 = 360

Why?
Ten 0.1s make a whole, so 1 ÷ 0.1 = 10
So dividing by 0.1 is the same as multiplying by 10.
One hundred 0.01s make a whole, so 1 ÷ 0.01 = 100
So dividing by 0.01 is the same as multiplying by 100.

Dividing by 0.1 makes each digit 10 times bigger.
Dividing by 0.01 makes each digit 100 times bigger.

(1) Work out

a 52 ÷ 0.1 = 620   **b** 257 ÷ 0.1 = 2570   **c** 3.1 ÷ 0.1 = 31

d 68.7 ÷ 0.1 = 687   **e** 9.02 ÷ 0.1 = 90.2   **f** 32.14 ÷ 0.1 = 324

(2) Work out

a 4 ÷ 0.01 = 400   **b** 38 ÷ 0.01 = 3800   **c** 51 ÷ 0.01 = 5100

d 2.9 ÷ 0.01 = 290   **e** 6.74 ÷ 0.01 = 00674   **f** 12.861 ÷ 0.01 = ~~0012861~~ 1286·1

(3) Work out

a 0.1 ÷ 0.1 = 1   **b** 9.01 ÷ 0.1 = 90.1   **c** 52.001 ÷ 0.01 = 5200.1

d 0.4 ÷ 0.1 = 4   **e** 10.01 ÷ 0.01 = 1001   **f** 0.5 ÷ 0.01 = 50

(4) Rashid divides a 5 kg bag of rice into 0.1 kg portions.
How many 0.1 kg portions of rice are there?   0.2

Exam-style question

(5) Work out

a 9 ÷ 0.1   90   (1 mark)   **b** 0.3 ÷ 0.01   (1 mark)

Reflect   How did you work out your answer to Q4? Did your work on Q1–3 help you?

Dividing negative integers

The rules for dividing integers are:
- positive ÷ positive = positive
- positive ÷ negative = negative
- negative ÷ positive = negative
- negative ÷ negative = positive

Guided practice

Work out

a $42 ÷ -7$ **b** $-3.52 ÷ 4$

---

a First work out the calculation without the signs.

$42 ÷ 7 =$ ....6........   Work out how many 7s go into 42.

Use the rules for dividing negative numbers.

positive ÷ negative = ...35....

Write the answer with the correct sign.

$42 ÷ -7 = -6$

b Work out the calculation without the signs.

$$\begin{array}{r} 0 . \\ 4\overline{)3 .\,^35\,\cdots2} \end{array}$$

Use the rules for dividing negative numbers.

negative ÷ positive = ...   Use the rule negative ÷ positive = negative and work out $3.52 ÷ 4$. Remember to line up the decimal points when dividing 3.52 by 4.

Write the answer with the correct sign.

$-3.52 ÷ 4 = -0.88$

---

(1) Work out

a $-100 ÷ 10 =$ ....−10....   **b** $-63 ÷ -7 =$ ....−70....   **c** $72 ÷ 12 =$ ....6....

d $64 ÷ -8 =$ ....56....   **e** $-48 ÷ 6 =$ ....−8....   **f** $24 ÷ -3 =$ ....21....

(2) Work out the missing numbers.

a $25 ÷ $ ....5.... $= -5$ **b** $-60 ÷$ .... $= 10$ **c** $-8 =$ .... $÷ 3$

d $-35 ÷$ .... $= -7$ **e** $-7 =$ .... $÷ 6$ **f** $-56 ÷ -8 =$ ....−64....

(3) Work out

a $-4.2 ÷ 0.1 =$ ....   **b** $-0.9 ÷ -0.01 =$ ....   **c** $5.32 ÷ -4 =$ ....

d $-12.54 ÷ 3 =$ ....   **e** $9.84 ÷ 8 =$ ....   **f** $-26.7 ÷ 4 =$ ....

Exam-style question

(4) Work out

a $12 ÷ -4$ .... **(1 mark)**   **b** $-0.5 ÷ -0.1$ .... **(1 mark)**

Reflect   Without looking at this page, write the rules for dividing positive and negative numbers.

Practise the methods

Answer this question to check where to start.

Check up

Tick the correct answer for 0.3 ÷ 0.01

A 0.003 ○   **B** 0.03 ○   **C** 0.3 ○   **D** 3 ○   **E** 30  ○   **F** 300 ○

If you ticked E go to Q2.	If you ticked A, B, C, D or F, go to Q1 for more practice.

1 Work out

a 4.1 ÷ 0.1
41

b 0.7 ÷ 0.1
7 0 7

c 0.5 ÷ 0.01
50

d 2.6 ÷ 0.01
260

Hint Dividing by 0.01 is the same as multiplying by 100.

Hint Dividing by 0.1 is the same as multiplying by 10.

	H	T	U	.	$\frac{1}{10}$
			4	.	1
× 10		4			

2 Work out

a 8.3 ÷ −0.1
8.2

b −2.5 ÷ 0.1
−25

c −5.4 ÷ −0.01
−5.41

d 9.1 ÷ −0.01
9.09

3 Work out

a 8.01 ÷ 3
2.67

b 7.68 ÷ 4
1.92

c 5.22 ÷ 9
0.58

d 6.44 ÷ 7
0.92

4 Work out

a −7.68 ÷ −3
10.68

b 69.2 ÷ −4
65.2

c −5.13 ÷ 9
0.57

d 48.6 ÷ 6
8.1

5 Work out

a 5.55 ÷ 2
2.775

b 32.8 ÷ 5
6.56

c 5.61 ÷ −4
1.61

d −56.7 ÷ −8
−64.7

Exam-style question

6 Work out

a 8 ÷ 0.1   *80*   (1 mark)

b 7.44 ÷ 6   *1.24*   (2 marks)

c −2.5 ÷ 4   *−0.625*   (3 marks)

Problem-solve!

1 Mr Carr needs to decide when to organise a school trip.

The table shows the number of students who will be on the trip on each of four days.

It also shows the number of teachers who can help on each of the four days.

	Tuesday	Wednesday	Thursday	Friday
Number of students	134	128	145	107 = 5 14
Number of teachers	9	7	8	6 = 30

For every 18 students on the trip there must be at least one teacher to help.

On which days will there be enough teachers to help on the trip?

You must show all your working.

$134 \div 18 = 7.4$ $145 \div 18 = 8.05$

$128 \div 18 = 7.1$ $107 \div 18 = 5.94$ $\div 18 =$

Tuesday and Friday

(3 marks)

2 Priya is organising a party for 126 adults and 48 children.

At 7 pm all the adults and all the children will sit down at tables for a meal.

Six people will sit at each table.

Work out the number of seats and the number of tables Priya will need.

$126 + 48 = 174$ 29 tables

$174 \div 6 = 29$ 174 chairs

Tables = 29 (3 marks)
Chairs = 174

3 Jodie has 159 eggs to put into boxes.

She can put six eggs into each box.

Find the smallest number of boxes Jodie needs.

You must show your working.

$159 \div 6 = 26.5$

27 (2 marks)

4 You can buy milk in two sizes of carton.

A 4 pint carton of milk costs £1.00.

A 6 pint carton of milk costs £1.47.

Which carton of milk is better value for money?

You must show all your working.

4 pints
£1.00

6 pints
£1.47

$£1.00 \div 4 = 0.25p$

$£1.47 \div 6 = 0.245p$

6 pints (3 marks)

Now that you have completed this unit, how confident do you feel?

1 Dividing 3-digit numbers by a single-digit number

2 Dividing decimals by a single-digit number

3 Dividing by 0.1 and 0.01

4 Dividing negative integers

③ Fractions

This unit will help you to use the four operations with fractions.

AO1 Fluency check

① Work out the missing numbers.

a $\dfrac{2}{2} = 1$ b $\dfrac{4}{4} = 1$ c $\dfrac{9}{9} = 1$ d $\dfrac{12}{12} = 1$

② Simplify

a $\dfrac{6}{10}$ *3/5*   b $\dfrac{10}{15}$ *2/3*   c $\dfrac{8}{20}$ ~~*4/10*~~ *2/5*

d $\dfrac{28}{35}$ *4/5*   e $\dfrac{27}{72}$ *3/8*   f $\dfrac{36}{84}$

③ Number sense

Write four multiples of

a 3   b 5   c 7   d 12

3,6,9,12   *5,10,15,20*   *7,14,21,28*   *12,24,36,48*

Key points

Equivalent fractions look different, but have the same value. For example, $\dfrac{1}{2}$ and $\dfrac{2}{4}$ are equivalent.

To add and subtract fractions, the denominators need to be the same.

These **skills boosts** will help you to convert between mixed numbers and improper fractions, add and subtract fractions and mixed numbers, multiply a fraction by an integer and divide by a fraction.

> **1** Mixed numbers and improper fractions
> **2** Adding and subtracting fractions and mixed numbers
> **3** Multiplying a fraction by an integer
> **4** Dividing an integer by a fraction

You might have already done some work on fractions. Before starting the first skills boost, rate your confidence using each concept.

① Write $4\dfrac{2}{3}$ as an improper fraction.

14/3

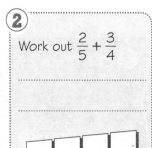

② Work out $\dfrac{2}{5} + \dfrac{3}{4}$

③ Work out $\dfrac{3}{7} \times 5$

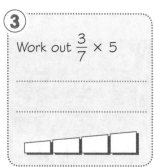

④ Work out $4 \div \dfrac{2}{5}$

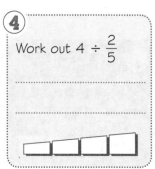

How confident are you?

1 Mixed numbers and improper fractions

An improper fraction is a fraction with the numerator greater than the denominator, e.g. $\frac{5}{3}$.

A mixed number has a whole-number part and a fraction part, e.g. $3\frac{1}{4}$.

When converting between mixed numbers and improper fractions, the denominator stays the same.

Guided practice

Worked exam question

a Write $3\frac{1}{4}$ as an improper fraction.

b Write $\frac{7}{3}$ as a mixed number.

---

a Use a bar model to help you.

Work out how many $\frac{1}{4}$s in 3.

$3\frac{1}{4} = \dfrac{\overset{3}{\ldots} \times 4 + 1}{4}$

$= \dfrac{13}{4}$

Use $\frac{4}{4} = 1$ to help you.

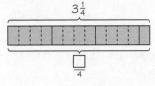

$3\frac{1}{4}$

$\frac{\square}{4}$

b Work out how many times the denominator goes into the numerator.

Use a bar model to help you.

$\frac{7}{3} = $

Use $\frac{3}{3} = 1$ to help you.

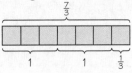

$\frac{7}{3}$

1   1 $\frac{1}{3}$

$7 \div 3 = \square$ remainder $\square$

---

① Write each mixed number as an improper fraction.

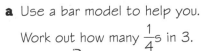

a $1\frac{1}{2} = $ 3/2   **b** $1\frac{2}{3} = $ 5/3   **c** $2\frac{1}{4} = $ 9/4   **d** $3\frac{1}{3} = $ 10/3

e $1\frac{3}{5} = $ 8/5   **f** $2\frac{4}{5} = $ 14/5   **g** $4\frac{1}{3} = $ 13/3   **h** $3\frac{3}{4} = $ 15/4

② Write each improper fraction as a mixed number.

a $\frac{5}{2} = $ 2½   **b** $\frac{4}{3} = $ 1⅓   **c** $\frac{7}{4} = $ 1¾   **d** $\frac{9}{5} = $ 1⅘

e $\frac{11}{4} = $ 2¾   **f** $\frac{8}{3} = $ 2⅔   **g** $\frac{12}{5} = $ 2⅖   **h** $\frac{12}{7} = $ 1 4/7

Exam-style question

③ **a** Write $2\frac{5}{8}$ as an improper fraction.   21/8   (1 mark)

b Write $\frac{23}{5}$ as a mixed number.   4 3/5   (1 mark)

Reflect

Can you explain how to write mixed numbers as improper fractions, and how to write improper fractions as mixed numbers, without using bar models?

② Adding and subtracting fractions and mixed numbers

When adding and subtracting fractions with different denominators, change the fractions to equivalent fractions that all have the same denominator.

Guided practice

Work out

a $\dfrac{7}{9} - \dfrac{5}{9}$ **b** $\dfrac{7}{8} + \dfrac{2}{3}$

a Subtract the numerators.

$\dfrac{7}{9} - \dfrac{5}{9} = \dfrac{7-5 \quad - \quad 5}{9}$

$= \dfrac{2}{9}$

b $\dfrac{7}{8} + \dfrac{2}{3}$

Find the lowest common multiple (LCM) of the denominators.

$= \dfrac{8}{24} + \dfrac{}{24}$

Add the numerators of the fractions.

$= \dfrac{}{24}$

Write as a mixed number.

$= 1\dfrac{13}{24}$

Use a bar model.

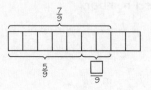

The LCM of 8 and 3 is 24.

① Work out

a $\dfrac{3}{5} + \dfrac{1}{5} = $ 4/5 ✓ **b** $\dfrac{5}{7} - \dfrac{2}{7} = $ 3/7 ✓ **c** $\dfrac{2}{9} + \dfrac{2}{9} = $ 4/9 ✓

② Work out

a $\dfrac{1}{2} + \dfrac{2}{7} = $ 11/14 **b** $\dfrac{2}{3} + \dfrac{1}{5} = $ 13/15 **c** $\dfrac{3}{10} + \dfrac{2}{5} = $ 7/10

③ Work out

a $\dfrac{5}{6} - \dfrac{1}{3} = $ 3/6 = 1/2 **b** $\dfrac{7}{8} - \dfrac{3}{5} = \dfrac{49}{40} = 1\dfrac{9}{40}$ **c** $\dfrac{7}{5} - \dfrac{3}{10} = \dfrac{37}{18}, 2\dfrac{1}{18}$

④ Work out   **Hint** Change your answer from an improper fraction to a mixed number.

a $\dfrac{3}{4} + \dfrac{2}{3} = $ .... **b** $\dfrac{5}{8} + \dfrac{3}{5} = $ .... **c** $\dfrac{11}{9} + \dfrac{5}{6} = $ ....

Exam-style question

⑤ Work out

a $\dfrac{1}{10} + \dfrac{1}{2}$ .... (2 marks) **b** $\dfrac{2}{3} - \dfrac{1}{5}$ .... (2 marks)

Reflect $\dfrac{3}{10} + \dfrac{2}{3} = \dfrac{5}{13}$ is wrong. Explain why.

3 Multiplying a fraction by an integer

When multiplying a fraction by an integer, multiply only the numerator by the integer. The denominator stays the same.

Guided practice

Work out $\frac{2}{3} \times 5$

Multiply the numerator by the integer.

$\frac{2}{3} \times 5 = \frac{2 \times 5}{3}$

$= \frac{10}{3}$

Write as a mixed number.

$= 3\frac{1}{3}$

Use a bar model to help you.

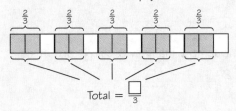

Total $= \frac{\square}{3}$

(1) Work out

a $\frac{1}{2} \times 10$

5

b $\frac{1}{3} \times 6$

2

c $\frac{1}{4} \times 20$

5

(2) Work out   **Hint** Simplify your answers.

a $\frac{3}{4} \times 16$

b $\frac{2}{3} \times 12$

c $\frac{4}{5} \times 5$

(3) Work out   **Hint** Write your answers as mixed numbers.

a $\frac{1}{4} \times 7$

b $\frac{3}{5} \times 4$

c $\frac{3}{7} \times 4$

(4) Work out   **Hint** Simplify your answers and write as mixed numbers.

a $\frac{5}{6} \times 3$

b $\frac{3}{4} \times 2$

c $\frac{7}{15} \times 3$

Exam-style question

(5) Work out $\frac{1}{4} \times 60$

.... (1 mark)

Reflect

$\frac{2}{5} \times 3 = \frac{6}{15} = \frac{2}{5}$ is wrong. Explain why.

4 Dividing an integer by a fraction

When dividing an integer by a fraction, work out how many times the fraction goes into the integer.

Guided practice

Work out

a $3 \div \frac{1}{4}$ **b** $4 \div \frac{2}{3}$ **c** $2 \div \frac{3}{5}$

---

a Work out how many quarters are in 3.

$3 \div \frac{1}{4} = $ ....

b Work out how many two-thirds are in 4.

$4 \div \frac{2}{3} = $ ....

c Work out how many three-fifths are in 2.

$2 \div \frac{3}{5} = $ .... $\dfrac{..............}{3}$

Alternatively, multiply 2 by the reciprocal of the fraction.

$2 \div \frac{3}{5} = 2 \times \frac{5}{3} = \dfrac{..............}{3} = ... \dfrac{..............}{3}$

Use a bar model to help you.

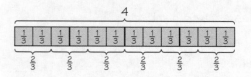

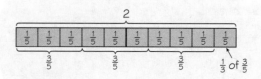

Why?

Dividing by a fraction is the same as multiplying by the reciprocal of the fraction.

The reciprocal of $\frac{3}{5}$ is $\frac{5}{3}$.

---

① Work out

a $3 \div \frac{1}{2}$ **b** $2 \div \frac{1}{4}$ **c** $2 \div \frac{1}{5}$

....

② Work out

a $2 \div \frac{2}{5}$ **b** $3 \div \frac{3}{4}$ **c** $4 \div \frac{2}{7}$

....

③ Work out

a $3 \div \frac{2}{3}$ **b** $3 \div \frac{4}{5}$ **c** $4 \div \frac{3}{4}$

....

Exam-style question

④ Work out $3 \div \frac{1}{5}$

.... (1 mark)

Reflect

Did you use bar models or the method using reciprocals to answer Q3? Go back and try the other method. Which method do you prefer, and why?

Practise the methods

Answer this question to check where to start.

Check up

Tick the correct calculation to work out $\frac{3}{10} + \frac{4}{5}$

A ◯
$$\frac{3}{10} + \frac{4}{5} = \frac{7}{15}$$

B ◯
$$\frac{3}{10} + \frac{4}{10} = \frac{7}{10}$$

C ◯
$$\frac{3}{10} + \frac{8}{10} = \frac{11}{10}$$

If you ticked C, change the improper fraction to a mixed number. Then go to Q4.

If you ticked A or B go to Q1 for more practice.

1 Work out the missing number in each equivalent fraction.

a $\frac{2}{3} = \frac{\ldots}{9}$

b $\frac{4}{5} = \frac{8}{\ldots}$

c $\frac{5}{12} = \frac{\ldots}{60}$

2 Work out

a $\frac{2}{7} + \frac{3}{7} = \frac{2 + 3}{7} = \frac{5}{7}$

b $\frac{5}{8} - \frac{3}{8} = \ldots$

c $\frac{5}{9} + \frac{2}{9} = \ldots$

3 Work out

a $\frac{1}{2} + \frac{1}{4} = \frac{\ldots + 1}{4} = \frac{2}{4}$

b $\frac{11}{12} - \frac{5}{6} = \ldots$

c $\frac{5}{8} + \frac{1}{4} = \ldots$

4 Work out

a $\frac{3}{4} + \frac{5}{6} = \ldots$

b $\frac{5}{9} + \frac{3}{5} = \ldots$

c $\frac{2}{3} + \frac{4}{5} = \ldots$

5 Work out

a $\frac{3}{4} \times 3 = \ldots$

b $5 \times \frac{5}{6} = \ldots$

c $\frac{5}{9} \times 3 = \ldots$

6 Work out

a $3 \div \frac{1}{2} = \ldots$

b $5 \div \frac{1}{3} = \ldots$

c $4 \div \frac{2}{5} = \ldots$

7 Work out

a $1 \div \frac{2}{5} = \ldots$

b $4 \div \frac{3}{5} = \ldots$

c $3 \div \frac{5}{6} = \ldots$

Exam-style question

8 Work out

a $\frac{1}{5} + \frac{3}{10}$ .... (2 marks)

b $\frac{3}{4} - \frac{1}{3}$ .... (2 marks)

c $\frac{1}{5} \times 40$ .... (1 mark)

d $5 \div \frac{1}{4}$ .... (1 mark)

9 Work out   **Hint** Multiply the numerators and multiply the denominators.

a $\frac{1}{2} \times \frac{1}{3} = \ldots$

b $\frac{3}{4} \times \frac{1}{2} = \ldots$

c $\frac{2}{3} \times \frac{1}{2} = \ldots$

Hint Multiply by the reciprocal of the second fraction.

d $\frac{1}{2} \div \frac{1}{2} = \ldots$

e $\frac{1}{2} \div \frac{1}{4} = \ldots$

f $\frac{2}{3} \div \frac{1}{6} = \ldots$

Problem-solve!

Exam-style questions

(1) James says that $\frac{7}{4\cancel{9}} - \frac{3}{\cancel{4}\cancel{9}} = \frac{4}{5}$ $\frac{28}{36} - \frac{27}{36} = \frac{1}{36}$

James is wrong. Explain why.

James needs to change the denominators and **(2 marks)** make them the same before taking the fractions away from eachother.

(2) There are 120 students in the school hall.

$\frac{1}{3}$ of the students leave the hall.

How many students are still in the hall?

33.3% of 120 = £39.96

120 ÷ 3

$\begin{array}{r} °\cancel{88}.\cancel{00} - \\ 39.96 \\ \hline 080.04 \end{array}$

80 students **(3 marks)**

(3) There are 40 counters in a bag.
The counters are green or yellow or blue.

$\frac{2}{5}$ of the counters are green.

$\frac{1}{4}$ of the counters are yellow.

$\begin{array}{r} 3\cancel{4}0 - \\ 22 \\ \hline 18 \end{array}$

Work out the number of blue counters in the bag.

40 ÷ 4 = 10 = yellow
30 ÷ 5 = 6×2 = 12 = green

18 blue counters **(3 marks)**

(4) ABCD is a square.

This is an accurate diagram.

What fraction of the square ABCD is shaded?

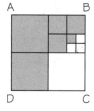

45/64 **(2 marks)**

75% of 25 = 18.75 $\overline{4}$

(5) **a** Hugh works out that $\frac{5}{8} \times 3 = \frac{5 \times 3}{8 \times 3} = \frac{15}{24}$

The answer $\frac{15}{24}$ is wrong.

Describe one mistake that Hugh has made.

Hugh is wrong because you can siplify **(1 mark)** it down to 5/8

b Work out the correct answer to $\frac{5}{8} \times 3$

Write your answer as a mixed number.

(2 marks)

Now that you have completed this unit, how confident do you feel?

1	2	3	4
Mixed numbers and improper fractions	Adding and subtracting fractions and mixed numbers	Multiplying a fraction by an integer	Dividing an integer by a fraction

④ Fractions, decimals and percentages

This unit will help you to convert between fractions, decimals and percentages, to write one number as a percentage of another, and to order and compare fractions, decimals and percentages.

AO1 Fluency check

① Work out
  a $1 \div 2 =$ *0.50* **b** $3 \div 4 =$ *0.75* **c** $3 \div 5 =$ *0.60*

  d $7 \div 10 =$ *0.70* **e** $3 \div 10 =$ *0.30* **f** $9 \div 5 =$ *1.80*

② **Number sense**

Complete the missing values in the table.

Fraction	Decimal	Percentage
$\frac{1}{2}$	0.5	*50%*
$\frac{1}{4}$	*0.25*	25%
1/3	0.75	*75%*

— Key points —

'Per cent' means 'out of 100', percentage means 'so many hundredths of'.

Fractions, decimals and percentages can be equivalent, e.g. $\frac{1}{2} = 0.5 = 50\%$

These **skills boosts** will help you to write one number as a percentage of another number, to convert between fractions, decimals and percentages, and to order and compare fractions, decimals and percentages.

① Converting between decimals and fractions	② Converting a fraction to a decimal	③ Writing one number as a percentage of another	④ Ordering and comparing fractions, decimals and percentages

You might have already done some work on fractions, decimals and percentages. Before starting the first skills boost, rate your confidence using each concept.

① Write 0.45 as a fraction. Give your answer in its simplest form.

② Write $\frac{26}{40}$ as a decimal.

③ Kelly achieved 62 out of 80 in her French test. Work out Kelly's result as a percentage.

④ Write these numbers in order of size. Start with the smallest.
0.6 $\frac{62}{100}$ 66%
0.606 $\frac{6}{100}$

How confident are you?

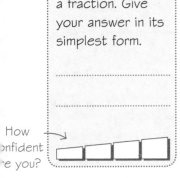
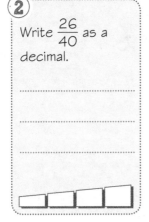
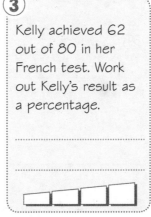
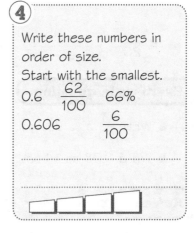

1 Converting between decimals and fractions

The place value of a decimal number tells you which fraction it is equivalent to.

Guided practice

a Write each decimal as a fraction, simplifying your answer where possible.
   i 0.3   **ii** 0.72

b Write $\frac{8}{25}$ as a decimal.

---

a **i** Write the decimal in a place-value table.
      Write the decimal as a fraction.

U	.	$\frac{1}{10}$
0	.	3

$$0.3 = \frac{\ldots\ldots}{10}$$

The lowest place value for 0.3 is tenths: $0.3 = \frac{\square}{10}$

   ii Write the decimal as a fraction.

U	.	$\frac{1}{10}$	$\frac{1}{100}$
0	.	7	2

$$0.72 = \frac{\ldots\ldots}{100}$$

Simplify the fraction.

$$= \frac{\ldots\ldots}{25}$$

The lowest place value for 0.72 is hundredths: $0.72 = \frac{\square}{100}$

b Write the equivalent fraction with a denominator of 100.

$$\frac{8}{25} = \frac{\ldots\ldots}{100}$$

Write the digits of the numerator as a decimal.

$$\frac{8}{25} = 0.32$$

U	.	$\frac{1}{10}$	$\frac{1}{100}$
0	.	......	......

$$\overset{\times 4}{\underset{\times 4}{\frac{8}{25} = \frac{\square}{100}}}$$

---

(1) Write each decimal as a fraction.

   a 0.23 ....   **b** 0.7 ....   **c** 0.67 ....   **d** 0.911 ....

(2) Write each decimal as a fraction. Simplify your answers.

   a 0.5 ....   **b** 0.8 ....   **c** 0.64 ....   **d** 0.08 ....

(3) Write each fraction as a decimal.

   a $\frac{1}{10}$ 0.2 ....   **b** $\frac{7}{10}$ 0 07 ....   **c** $\frac{1}{5}$ 0.5 ....   **d** $\frac{3}{5}$ ....

   Hint Use equivalent fractions so that the denominator is 10 or 100.

   e $\frac{1}{20}$ ....   **f** $\frac{9}{20}$ ....   **g** $\frac{17}{20}$ ....   **h** $\frac{24}{25}$ ....

Exam-style question

(4) **a** Write $\frac{3}{10}$ as a decimal.   .... **(1 mark)**

   b Write 0.08 as a fraction.   .... **(1 mark)**

Reflect Explain how to use $\frac{1}{25} = 0.04$ to write $\frac{11}{25}$ as a decimal.

2 Converting a fraction to a decimal

You can use division to convert a fraction to a decimal: to write $\frac{3}{16}$ as a decimal, work out $3 \div 16$.

Guided practice

Write $\frac{11}{16}$ as a decimal.

Worked exam question

Divide the numerator by the denominator.

$$16 \overline{)11.0^{14}0^{12}0\cdots0} \quad \begin{array}{c} 0.6\ 8 \\ \end{array}$$

$\frac{11}{16} = $ ....

Work out $11 \div 16$.
Write a decimal point and zeros to show the rest of the calculation.
Make sure you line up the decimal points.
You may need to write the multiples of 16:
16, 32, 48, 64, 80, 96, 112, 128, 144.

① Write each fraction as a decimal.

a $\frac{3}{20}$ *0.15*

b $\frac{14}{25}$ *0.56*

c $\frac{4}{5}$ *0.8*

d $\frac{7}{40}$ *0.175*

e $\frac{23}{40}$ *0.575*

f $\frac{19}{80}$ *0.2375*

② Write each fraction as a decimal.

a $\frac{5}{8}$ *0.625*

b $\frac{3}{16}$ *0.1875*

c $\frac{25}{32}$ *0.78125*

d $\frac{7}{16}$ *0.4375*

e $\frac{27}{40}$ *0.675*

f $\frac{57}{80}$ *0.7125*

③ Write each improper fraction as a decimal.

Hint Use the same method as for proper fractions.

a $\frac{8}{5}$ *1 3/5*

b $\frac{27}{20}$ *1 7/20*

c $\frac{57}{40}$ *1 17/40*

d $\frac{32}{25}$ *1 7/25*

e $\frac{21}{8}$ *2 5/8*

f $\frac{201}{80}$ *2 41/80*

Exam-style question

④ Write $\frac{9}{16}$ as a decimal.

0.5625 (1 mark)

Reflect   All of your answers to Q1 and Q2 started with a zero before the decimal point.
Explain why your answers to Q3 didn't start with a zero.

3 Writing one number as a percentage of another

To write one number as a percentage of another, first use the two numbers to write a fraction, then convert your fraction to a percentage: 9 out of 20 = $\frac{9}{20}$ = 45%.

Guided practice

Kyle gets 37 out of 40 in a Maths test.
Work out Kyle's result as a percentage.

Write 37 out of 40 as a fraction.

37 out of 40 = $\frac{37}{40}$

Put the number that the score is 'out of' as the denominator.

Use division to convert the fraction to a decimal.

$$0.9\,4$$
$$40\overline{)37.0^{10}0\cdots0}$$

$\frac{37}{40}$ = ...

Multiples of 40:
40, 80, 120, 160, 200, 240, 280, 320, 360, 400

Multiply your decimal by 100 to write it as a percentage.

.... × 100 = 92.5%

(1) Work out

a 13 as a percentage of 20

b 18 as a percentage of 25

c 43 as a percentage of 50

d 13 as a percentage of 16

(2) Work out these test results as percentages.

a 19 out of 20   **b** 3 out of 8   **c** 16 out of 25   **d** 62 out of 80

Exam-style question

(3) There are 40 pens in a box. 18 of the pens are green.
Work out the percentage of pens that are green. .... **(2 marks)**

(4) Write 50p as a percentage of £5.   **Hint** Change both amounts to pence. £1 = 100p.

(5) Write 30 cm as a percentage of 2 metres.

(6) Write 750 g as a percentage of 3 kg.   **Hint** 1000 g = 1 kg

(7) Write £5.60 as a percentage of £8.

Reflect Compare any different methods you used on this page. Which do you prefer and why?

4 Ordering and comparing fractions, decimals and percentages

When ordering and comparing fractions, decimals and percentages, change everything to either all decimals or all percentages to make the comparisons easier.

Guided practice

Write these numbers in order of size, starting with the smallest.

$\dfrac{11}{20}$ 0.52 $\dfrac{1}{5}$ 0.505   53%

Change the fractions and decimals to percentages.

$\dfrac{11}{20} = 20 \overline{\smash{)}11.0^{10}0}^{0.5\ldots} = \ldots\ldots \times 100 = \ldots\ldots\%$

$0.52 \times 100 = \ldots\ldots\%$ $\dfrac{1}{5} = \ldots\ldots\%$ $0.505 \times 100 = \ldots\ldots\%$

Compare the percentages to put the numbers in order.

Check whether the question asks for the numbers to be ordered smallest first, or largest first.

When writing your final answer, use the original fractions, decimals and percentages from the question.

$\dfrac{1}{5}$, $\ldots\ldots$, 0.52, $\ldots\ldots$, $\dfrac{11}{20}$

① Write these numbers in order of size, starting with the smallest.

0.75   25% $\dfrac{1}{2}$ 0.3   100%

....

② Write these numbers in order of size. Start with the smallest number.

0.9   80% $\dfrac{3}{4}$ 0.35 $\dfrac{2}{5}$

....

③ Here are five numbers.

45% $\dfrac{1}{4}$ 0.4   0.404   44%

Hint 'Descending' means from largest to smallest.

Write these numbers in descending order.

....

④ A survey is carried out in two schools to compare the proportion of students who walk to school.
In school A, 63% of students walk to school.
In school B, $\dfrac{7}{10}$ of students walk to school.
Which school has the larger proportion of students who walk to school? ....

Exam-style question

⑤ Here are four numbers.   0.37 $\dfrac{3}{8}$ 36% $\dfrac{7}{20}$

Write these numbers in order of size.
Start with the smallest number. ....   **(2 marks)**

Reflect   Look at your working for each question on this page. Did you change the numbers to all decimals or all percentages? Which is the easiest to change, and why?

Practise the methods

Answer this question to check where to start.

Check up

Tick the correct decimal equivalent to $\frac{1}{5}$.

 A 0.5 ✓   **B** 0.05 ◯   **C** 1.5 ◯   **D** 0.2 ◯

If you ticked D go to Q2. | If you ticked A, B or C go to Q1 for more practice.

 1 Match the equivalent fractions and decimals.

$$\frac{3}{5} \qquad \frac{3}{10} \qquad \frac{3}{20} \qquad \frac{9}{10} \qquad \frac{4}{5} \qquad \frac{19}{20} \qquad \frac{18}{25} \qquad \frac{7}{10}$$

$$\frac{15}{100} \qquad \frac{95}{100} \qquad \frac{72}{100} \qquad \frac{60}{100} \qquad \frac{30}{100} \qquad \frac{70}{100} \qquad \frac{90}{100} \qquad \frac{80}{100}$$

0.95   0.7   0.9   0.8   0.15   0.6   0.3   0.72

2 Write each decimal as a fraction in its simplest form.

   a 0.16   **b** 0.92   **c** 0.55   **d** 0.42

3 Write each fraction as a decimal.

   a $\frac{3}{8}$ **b** $\frac{5}{16}$ **c** $\frac{7}{40}$ **d** $\frac{15}{16}$

4 Work out

   a 41 as a percentage of 50   **b** 22 as a percentage of 25

   c 11 as a percentage of 20   **d** 15 as a percentage of 32

5 Write £3.50 as a percentage of £8.

Exam-style question

6 Here are four numbers.   0.37 $\frac{3}{8}$ 36% $\frac{7}{20}$

   Write these numbers in order of size.

   Start with the smallest number.   (2 marks)

Problem-solve!

(1) Complete the missing values in the table.

Fractions	$\frac{1}{3}$	$\frac{2}{3}$	$\frac{1}{9}$	....	$\frac{5}{9}$	$\frac{7}{9}$	....
Decimals	0.33333...	....	0.11111...	0.22222...	....	....	0.44444...

Exam-style questions

(2) Work out the difference in value between $\frac{1}{4}$ and 30%. .... (2 marks)

(3) Work out £1.25 as a percentage of £8. .... (2 marks)

(4) Jake wants to make two fruit cakes.
He weighs some dried fruit.
The dried fruit weighs 0.9 kg.
Jake needs $\frac{3}{4}$ kg of dried fruit to make one cake.
Work out how much more dried fruit Jake needs to make two cakes.
Give your answer in grams. .... (4 marks)

(5) Amy, Briony and Ceri did a test.
The total for the test was 80 marks.
Amy got 56% of the marks.
Briony got $\frac{9}{16}$ of the marks.
Ceri got 44 out of 80.
Who performed the best in the test?
You must show all your working. .... (3 marks)

(6) In one week, a sandwich bar sells 80 baguettes out of a total of 250 sales.
Work out the percentage of the sales that were baguettes.

.... (2 marks)

(7) Here are four numbers.   0.67 $\frac{2}{3}$ 65.5% $\frac{11}{16}$
Write these numbers in order of size.
Start with the smallest number. .... (2 marks)

Now that you have completed this unit, how confident do you feel?

1 Converting between decimals and fractions

2 Converting a fraction to a decimal

3 Writing one number as a percentage of another

4 Ordering and comparing fractions, decimals and percentages

⑤ Probability

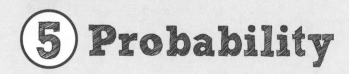

This unit will help you work out probability and use experimental probability and frequency trees.

AO1 Fluency check

① There are 20 marbles in a bag. 9 of the marbles are red, 7 are green and the rest are blue.

Write down the fraction of the marbles that are green. ....

② In a school, 70% of the students live in the catchment area.
Write down the percentage of students who do *not* live in the catchment area. ....

③ **Number sense**

Work out

a $1 - 0.6 =$ .... **b** $1 - 0.2 =$ .... **c** $1 - \frac{1}{4} =$ .... **d** $1 - \frac{2}{5} =$ ....

Key points

| Probability means the chance of an event happening. | The probability of an outcome = $\dfrac{\text{number of ways the outcome can happen}}{\text{total number of possible outcomes}}$ |

These **skills boosts** will help you to use the probability scale, understand mutually exclusive outcomes, predict the number of successes for an experiment, and draw and use frequency trees.

| ① The probability scale | ② Mutually exclusive outcomes for one event | ③ Estimating successes | Mutually exclusive outcomes for two events and frequency trees |

You might have already done some work on probability. Before starting the first skills boost, rate your confidence using each concept.

① The probability that it will rain tomorrow is $\frac{4}{7}$. What is the probability that it will **not** rain tomorrow?

② A bag contains 4 red, 2 blue and 3 yellow marbles. Work out the probability of randomly picking a blue marble from the bag.

③ The probability that a spinner lands on an even number is $\frac{3}{5}$. The spinner is spun 600 times. Work out how many times the spinner is expected to land on an even number.

④ A new car is sold in black, red or silver, and in three different models: saloon, estate and hatchback. List all possible combinations for the new car.

How confident are you?

1 The probability scale

The probability scale is labelled from 0 to 1 and is used to position and compare probabilities.
The sum of the probabilities of mutually exclusive outcomes is 1.

Guided practice

The probability of this spinner landing on 3 is $\frac{1}{3}$.

On a probability scale, mark with a cross the probability of the spinner

a landing on 3　　　**b** not landing on 3.

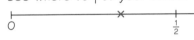

a $P(3) = \frac{1}{3}$

P(3) means 'the probability of a 3'.

Divide the probability scale into thirds to see where to put your cross.

0 —×———— $\frac{1}{2}$ ———— 1

Use your ruler to measure the scale and then divide it by 3.

b $P(3) = \frac{1}{3}$

Use the fact that the probabilities of mutually exclusive outcomes add up to 1.

Mutually exclusive events are outcomes that cannot happen at the same time.

$P(\text{not } 3) = 1 - P(3) = 1 - \frac{\ldots}{\ldots} = \frac{\ldots}{\ldots}$

Mark the calculated probability on the probability scale.

0 ————————×— $\frac{1}{2}$ ———— 1

---

① Mark with a cross the probability of each of these events.
　a The probability that it will rain tomorrow is 60%.
　b The probability that a biased dice lands on 1 is 0.25.
　c The probability of winning a tennis match is 0.7.
　d The probability that a day of the week selected at random contains the letter 'y'.

Hint Convert the percentage to a decimal.

0 ———————— 0.5 ———————— 1

② The probability of picking a red counter out of a bag is 0.3.
Work out the probability of picking a counter that is *not* red.

Hint
P(not red) = 1 − P(red) = 1 − 0.3

③ The probability of a spinner landing on red is $\frac{4}{9}$.

Work out the probability of the spinner *not* landing on red.

Exam-style question

④ Some male and some female students are attending tennis coaching.

The probability of picking a male at random for a match is $\frac{1}{3}$.

What is the probability of picking a female?　　　　　　　　　(1 mark)

Reflect　When you know the probability of an event happening, how do you work out the probability of the event *not* happening?

2 Mutually exclusive outcomes for one event

Mutually exclusive events are events that cannot happen at the same time. For example, getting a head or tail when flipping a coin are mutually exclusive outcomes because the coin can only land on one or the other. Getting a queen or a heart when dealing a pack of cards are not mutually exclusive outcomes because the card could be both outcomes, the queen of hearts.

Guided practice

Worked exam question

A letter is picked at random from the words MUTUALLY EXCLUSIVE.

Work out the probability that the letter is a vowel.

_____

Number of ways of getting a vowel = ...

Number of possible outcomes = ...

Number of possible outcomes = number of letters

Work out the probability.

Probability of a vowel = $\frac{7}{17}$

Probability of a vowel = $\frac{\text{number of vowels}}{\text{total number of letters}}$

① Write whether each pair of outcomes are mutually exclusive or not.

Hint Mutually exclusive events are outcomes that cannot happen at the same time.

   a Spinning a 5-sided spinner, labelled A to E, the result is a vowel or a consonant.

....

   b Dealing an ace or a diamond from a pack of cards.

....

   c Rolling an even or an odd number with a 6-sided dice.

....

② Here are 10 letters.　　A　　A　　B　　C　　C　　C　　D　　E　　E　　F

Amir takes a letter at random.

   a Write down the probability that he takes a letter C.

....

   b Write down the probability that he does *not* take a letter C.

....

③ A bag contains 7 white beads and 13 red beads.
   A bead is taken at random from the bag.

   a Write down the probability that the bead is white.

....

   b Work out the probability that the bead is red.

....

Exam-style question

④ There are 110 cars in a college car park. Teachers own 71 of the cars. Students own 17 of the cars. One of the cars in the car park is picked at random.

   a Write the probability that a teacher owns this car.

.... (1 mark)

   b Work out the probability that the car is *not* owned by either a teacher or a student.

.... (3 marks)

Reflect　　What is the total of the probabilities of all the mutually exclusive outcomes to an event? Use your answers to Q2 to help you.

3 Estimating successes

To estimate the number of successes, multiply the probability of a success by the number of trials.

Guided practice

The probability of a spinner landing on red is $\frac{2}{5}$, on blue is 0.5 and on green is 10%.
Callum spins the spinner 300 times.
Estimate the number of times the spinner lands on

a red   **b** blue   **c** green.

---

a Multiply the probability of landing on red by the number of spins.
Estimate for the spinner landing on
red = $\frac{2}{5}$ × 300 = ... times

b Estimate for the spinner landing on
blue = 0.5 × 300 = ... times

c Estimate for the spinner landing on
green = 10% of 300 = ... times

$\frac{2}{5}$ × 300 = $\frac{2}{5}$ of 300
Use a bar model.

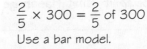

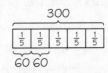

---

① Here is a 4-sided biased spinner.
The table shows the probability for each outcome.

Number	1	2	3	4
Probability	0.1	0.3	0.2	0.4

The spinner is spun 200 times.
Estimate the number of times the spinner lands on 2. ....

② Erin has a biased coin. She flips the coin once.
The probability of getting heads is 0.3.
a Work out the probability of getting tails. ....

Jamal flips this coin 200 times.
b Estimate the number of tails Jamal gets. ....

③ Toby plays a game. The probability that Toby wins the game is 25%.
Toby plays the game 240 times.
Work out an estimate for the number of times he wins the game. ....

④ An ordinary 6-sided dice is rolled in a game. Bonus points are scored for rolling a 6.
During the game, the dice is rolled 120 times.
Estimate the number of times the dice lands on 6. ....

Exam-style question

⑤ Oti takes a counter from a bag at random.
The probability that she takes a red counter is 40%.
Oti writes down the colour of the counter and returns it to the bag. Oti does this 50 times.
Work out an estimate for the number of times that Oti takes
a red counter from the bag. .... **(2 marks)**

Reflect   Explain how you know that the spinner in Q1 is biased.

4 Mutually exclusive outcomes for two events and frequency trees

When listing outcomes for more than one event, be systematic.
You can use sample space diagrams and two-way tables to list outcomes for two or more events.
Frequency trees show the number of options for different outcomes.

Guided practice

Worked exam question

Hasan flips two coins once.
Each coin will land on either heads or tails.

a List all the possible outcomes that Hasan could get.

b Work out the probability that Hasan gets two tails.

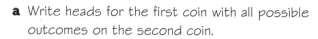

a Write heads for the first coin with all possible outcomes on the second coin.

Coin 1	Coin 2
heads	heads
heads	

Now write tails for the first coin with all possible outcomes on the second coin.

....

....

> Be systematic with your list to help cover all possible outcomes.

b Write the number of possible outcomes.

Number of possible outcomes = ...

Probability of two tails = $\dfrac{1}{\rule{2cm}{0.4pt}}$

> Count the number of possible outcomes in part **a**.

> Probability of an outcome = $\dfrac{\text{number of ways the outcome can happen}}{\text{total number of possible outcomes}}$

1 A café sells a soup and sandwich on a special offer.
Customers can choose tomato or vegetable soup and a ham or cheese sandwich.
List all possible combinations of the soup and sandwich.

....

2 Six students stand for school council.
Three of the students are boys: Alfie, Tom and George.
Three of the students are girls: Lexi, Clara and Esme.
One boy and one girl are picked at random.
List all possible outcomes for the school council.

....

3 Here are two 3-sided fair spinners.

Zainub is going to spin each spinner once.

Her score is the sum of the two numbers.

Spinner A   Spinner B

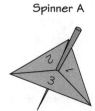

a Complete the sample space diagram for each possible score.

Spinner A

		1	2	3
Spinner B	1	2	3	4
	2	....	....	....
	3	....	....	....

b Work out the probability that Zainub gets

   i a score of 4 ....

   ii a score greater than 3. ....

④ A mobile phone company surveys 200 of its customers.
162 of these customers have a smartphone.
154 of the 200 customers have a contract.
27 of the 'other mobile phone' customers use
'pay as you go'.

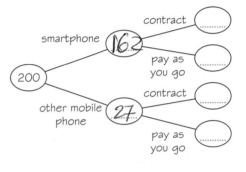

  a Use this information to complete the frequency tree.

One of the smartphone customers is selected at random.

  b Work out the probability that this smartphone customer
has a contract.

....

  c Complete the two-way table using the information given in this
question.

Hint
Put the numbers you
are given into the
correct cells in the
table. Next work out
missing numbers in
rows and columns
where you have two
values.

	Smartphone	Other mobile phone	Total
Contract	....	....	....
Pay as you go	....	....	....
Total	....	....	....

Exam-style question

⑤ 120 students had some homework to do.
52 of the students were boys.
15 of the 120 students did *not* do their
homework.
63 of the girls did do their homework.

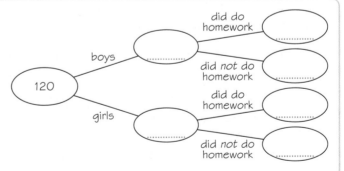

  a Use this information to complete the frequency tree. **(3 marks)**

One of the girls is selected at random.

  b Work out the probability that this girl did not do
her homework. .... **(2 marks)**

Reflect In Q4, did you find the frequency tree or the two-way table easier for displaying the
data? Why?

Practise the methods

Answer this question to check where to start.

(1) Work out

a 1 − 0.3 = ...

b 1 − 0.65 = ...

c 1 − 0.25 = ...

d 1 − 0.83 = ...

(2) The probability that a spinner lands on red is 0.45.
Work out the probability that it does *not* land on red. ....

(3) A packet of sweets contains 5 orange and 7 red sweets.
A sweet is taken from the packet at random.

a Write down the probability that the sweet is orange. ....

b Work out the probability that the sweet is red. ....

(4) In a game, there are cards with the numbers 1 to 4.
The table shows the probability of getting each number.

Number	1	2	3	4
Probability	0.24	0.4	0.16	

a Work out the probability of getting a 4. ....

b The cards are shuffled 160 times during the game and each time a
card is dealt. Estimate the number of times a 2 is dealt. ....

(5) Taylor takes a letter tile from a bag at random. The probability that she takes a vowel is 30%.
Taylor writes down the letter on the tile. She then puts the tile back in the bag.
Taylor does this 80 times.
Estimate the number of times that Taylor takes a vowel. ....

Exam-style question

(6) Here is the menu in a café.

Sally is having a meal in the café.

She chooses one starter and one main course.

List all the different meals Sally can choose.

Menu	
Starter	**Main course**
Melon	Pasta
Soup	Chicken
	Fish

.... (2 marks)

Problem-solve!

1. There are 10 sweets in a box. x sweets are red. The rest of the sweets are yellow.
   Tracy takes a sweet from the box at random.
   Write down an expression, in terms of x, for the probability
   that Tracy takes a yellow sweet. .... (2 marks)

2. Kate has a 4-sided spinner.
   The sides of the spinner are numbered 1, 2, 3 and 4.
   The spinner is biased.
   The table shows the probability that the spinner lands on 1, 3 or 4.
   The probability that the spinner lands on 3 is x.

Number	1	2	3	4
Probability	0.2		x	0.1

   a Find an expression, in terms of x, for the probability that the
   spinner lands on 2. Give your answer in its simplest form. .... (2 marks)

   Kate spins the spinner 300 times.

   b Write down an expression, in terms of x, for the number of
   times the spinner is likely to land on 3. .... (1 mark)

3. A restaurant sold 180 pizzas last week.
   103 of the pizzas had a thin crust base.
   49 of the deep pan bases had a meat topping.
   64 of all the pizzas had a vegetarian topping.

   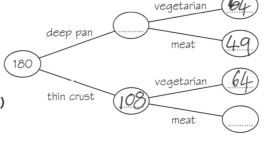

   a Use this information to complete the
   frequency tree. **(3 marks)**

   One of the customers who chose a deep pan base
   is selected at random.

   b Work out the probability that this customer had a meat
   topping on their pizza. .... (2 marks)

4. Lei has a biased coin. When she flips the coin once, the probability of getting heads is x.

   a Write down an expression, in terms of x, for the probability
   of getting tails. .... (1 mark)

   Lei flips the coin 100 times.

   b Write down an expression, in terms of x, for an estimate of
   the number of times she gets heads. .... (2 marks)

Now that you have completed this unit, how confident do you feel?

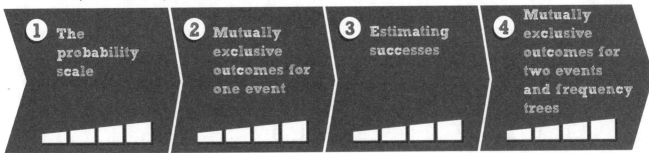

1 The probability scale

2 Mutually exclusive outcomes for one event

3 Estimating successes

4 Mutually exclusive outcomes for two events and frequency trees

Ratio and proportion

This unit will help you to use and find ratios, use the unitary method to solve problems and solve inverse proportion problems.

A01 Fluency check

① List all the factors of

a 10

1, 2, 10, 5

b 12

1, 12, 2, 6

c 25

1, 25, 5,

d 24

1, 2, 4, 12, 24

e 35

35, 5, 7

f 50

1, 50, 10, 5

② Write the highest common factor (HCF) of each pair of numbers.

a 12 and 20

b 27 and 45

c 48 and 120

③ **Number sense**

Simplify each fraction.

a $\frac{6}{10}$ *1/3*

b $\frac{15}{25}$ *3/5*

c $\frac{9}{21}$ *3/7*

d $\frac{24}{40}$ *6/10 3/5*

e $\frac{35}{42}$

f $\frac{60}{80}$

Key points

Ratio is used to compare amounts to each other.

Proportion is used to compare a part with the total amount.

These **skills boosts** will help you to use and find ratios, and solve proportion problems.

1 Simplifying and using ratio

2 Using proportion

3 Inverse proportion

You might have already done some work on ratio. Before starting the first skills boost, rate your confidence using each concept.

① Jack and Abby share some sweets in the ratio 4 : 3. Jack gets 36 sweets. How many sweets does Abby get?

② 5 drinks cost £4.15. How much do 8 drinks cost?

/

③ Two builders take 5 hours to build a wall. How long does it take four builders to build a wall of the same size?

How confident are you?

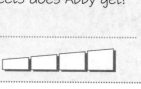

① Simplifying and using ratio

To simplify a ratio, divide the numbers in the ratio by their highest common factor (HCF).

Guided practice

a In a maths group there are 18 boys and 12 girls.
Write down the ratio of boys to girls.
Give your ratio in its simplest form.

b In Year 11 the ratio of boys to girls is 4 : 3.
There are 128 boys in Year 11.
Work out the number of girls in Year 11.

---

a Write the ratio using the numbers in the question.

_18 : 12_

Write the ratio in its simplest form.
Divide both sides by the highest common factor (HCF).
= 3 : 2

$$\div 6 \left(\begin{array}{c} 18 : 12 \\ \square : \square \end{array} \right) \div 6$$

The HCF of 18 and 12 is 6.

b Work out how many students are in 1 'part'.
4 parts = 128 students

1 part = 128 ÷ _12_ = ...

Multiply the number in 1 part by the number
of parts represented by girls.

3 parts = ... × 3

There are .... girls in Year 11.

Boys : Girls

$$\times \square \left(\begin{array}{c} 4 : 3 \\ 128 : \square \end{array} \right) \times \square$$

---

① Simplify each ratio.

a 8 : 10 = ~~4.5~~   **b** 4 : 12 = ...   **c** 25 : 20 = ...   **d** 32 : 24 = ...

e 21 : 35 = ...   **f** 16 : 36 = ...   **g** 144 : 60 = ...   **h** 63 : 18 = ...

② Ewan has 48 white tiles and 16 blue tiles.
Write down the ratio of white tiles to blue tiles.
Give your answer in its simplest form.   ....

③ Alice and Beth share some money
in the ratio 2 : 3.
Alice gets £26.
How much does Beth get?   ....

Hint   A : B

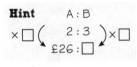

$$\times \square \left(\begin{array}{c} 2 : 3 \\ £26 : \square \end{array} \right) \times \square$$

④ The ratio of sugar to butter in a cake recipe is 2 : 5.
John makes a cake using 150 g of sugar.
How many grams of butter does John use?   ....

Exam-style question

⑤ The ratio of the number of males to the number of females in a running club is 8 : 7.
There are 91 females in the running club.
How many males are there in the running club?   ....   **(2 marks)**

2 Using proportion

In the unitary method, first work out the value of one, and then use this to work out the value of more.

Guided practice

6 identical pens cost £9.
How much do 10 of these pens cost?

Work out the cost of 1 pen.

1 pen costs £9 ÷ 6 = £ *1.60*

Multiply the cost of 1 pen by 10.

10 pens cost £.................... × 10

= £15

You can use a bar model to help you:

Alternatively, use an arrow diagram:

	Pens	Cost	
÷6 (	6	£9	) ÷6
×10 (	1	£☐	) ×10
	10	£☐	

① 3 kitchen chairs cost a total of £74.91.
How much do 8 chairs cost?

② Jack needs 8.6 m of ribbon for 4 banners.
How much ribbon does he need for 7 banners?

③ 5 jars weigh 2.275 kg. How much do 12 jars weigh?

Exam-style question

④ 3 calculators cost £35.85.
How much do 5 calculators cost? .... (3 marks)

⑤ The exchange rate for US dollars is £1 = $1.20.
How many dollars can you buy with £300? ....

Hint

	£	$
×☐ (	1	1.2	) ×☐
	300	☐	

⑥ Ahmed is going on holiday to Turkey.
The exchange rate is £1 = 3.6383 lira.
Ahmed changes £450 to lira. How many lira does he get?

⑦ 5 miles is the same distance as 8 km.
Work out how many kilometres there are in 12.5 miles.

⑧ The ratio of the number of boys to the number of girls in a class is 3 : 4.
What fraction of the class is boys?

Hint

| B | B | B | G | G | G | G |

How many students in total?

⑨ In a school, 40% of the staffing budget is spent on support staff.
The rest of the staffing budget is spent on teachers.

Hint Use an arrow diagram. n does not have to be a whole number.

  a What is the ratio of the spending on support staff to the spending on teachers? Give your answer in the form 1 : n.

The ratio of the number of staff to the number of students is 1 : 8.

  b What fraction of the school is students?

Reflect

Which method did you use for Q1 to Q4: arrow diagram, bar model or another method?
Go back and try an alternative method. Which method do you find easier?

3 Inverse proportion

Inverse proportion is when one value increases at the same rate as the other value decreases.

Guided practice

2 builders build a wall in 12 hours.
How long will it take 3 builders to build an identical wall?
Give your answer in hours and minutes.

Work out how long it would take 1 builder.
2 builders take 12 hours.

1 builder takes 12 × ....2.... = ... hours.

3 builders take .... ÷ 3 = 8 hours.

Builders	Hours
2	12
1	1
3	3

÷2 (...) ×2 ← inverse of ÷2
×3 (...) ÷3 ← inverse of ×3

① It takes 2 painters 5 days to paint a house.

   a How many days will it take 1 painter to paint an identical house? ....

   b How many days will it take 5 painters to paint an identical house? ....

② 4 plasterers plaster a house in a day.
How many days will it take 2 plasterers to plaster a similar house? ....

③ 2 roofers tile a roof in 10 hours.
How long will it take 5 roofers to tile a similar roof? ....

④ A gardener plants a border in half an hour.
How long will it take 5 gardeners to plant a similar border?
Give your answer in minutes. ....

⑤ A farmer has enough food to feed 200 goats for 9 days.
The farmer goes to market and buys 100 more goats.
For how many days will the food last? ....

⑥ 2 gardeners can plant a box of bulbs in $1\frac{1}{2}$ hours.
How many gardeners will it take to plant a similar box of bulbs
in 30 minutes? ....

⑦ 3 darts players throw some darts in 10 minutes.
How long will it take 9 darts players to throw
the same number of darts?
Give your answer in minutes and seconds.

Hint Multiply the decimal by 60 seconds, e.g.
0.5 minute = 0.5 × 60 = 30 seconds.

....

Exam-style question

⑧ In a factory it takes 5 machines 6 hours to fill a batch of yogurt pots.
The company needs to have one machine serviced.
How long will it take the remaining machines to fill a batch of yogurt pots? ....

(2 marks)

Reflect

Use your workings to these questions to explain inverse proportion.

Practise the methods

Answer this question to check where to start.

Check up

Tick the correct fully simplified ratio for 8 : 20.

Ⓐ 2 : 5 ○   Ⓑ 8 : 20 ○   Ⓒ 16 : 40 ○   Ⓓ 1 : 2.5 ✓   Ⓔ 4 : 10 ○

| If you ticked A go to Q3. | If you ticked B, C, D or E go to Q1 for more practice. |

① **a** Work out the HCF of 8 and 20.

....

b Simplify the ratio 8 : 20 using your answer to part **a**.

....

② **a** Work out the HCF of 18 and 30.

....

b Simplify the ratio 18 : 30 using your answer to part **a**.

....

③ Samia has 24 romantic comedy DVDs and 32 action DVDs.
Write down the ratio of romantic comedy DVDs to action DVDs.
Give your answer in its simplest form.   ....

④ The ratio of track athletes to field athletes in a competition is 6 : 5.
There are 102 track athletes.
How many field athletes are in the competition?   ....

⑤ 5 pizzas cost £42.45.
How much do 8 of these pizzas cost?   ....

⑥ Zach is going on holiday to New York.
The exchange rate is £1 = $1.20.
Zach changes £580 to dollars.
How many dollars does he get?   ....

⑦ 30 cm is approximately equivalent to 12 inches.
Dave is 66 inches tall. Work out his height in cm.   ....

⑧ The ratio of squash to water in a drink is 2 : 5.
What fraction of the drink is water?   ....

Exam-style question

⑨ Write the ratio 5 : 8 in the form 1 : n.

.... **(1 mark)**

⑩ It will take 2 authors 9 weeks to write a textbook.
How long will it take 5 authors to write the same textbook?   ....

Problem-solve!

1 Simplify

   a 60 cm : 1.4 m   **b** 250 g : 1.2 kg   **c** 55 mm : 4 cm   **d** 350 ml : 0.6 litres

....

Exam-style questions

2 5 schools sent some students to a conference.
One of the schools sent both boys and girls.
The ratio of the number of boys it sent to the number of girls it sent was 1 : 2.
It sent 12 boys. The other 4 schools sent only girls.
Each school sent the same number of students.
How many students in total were sent to the conference
by these 5 schools?   .... **(4 marks)**

3 A pack of cards has 48 cards. Each card is either blue or green.
The ratio of the number of blue cards to the number of green cards is 1 : 1.
8 blue cards are removed from the pack.
What is the new ratio of the number of blue cards to the number of
green cards? Give your answer in its simplest form.   .... **(3 marks)**

4 The total weight of 3 tins of beans and 4 jars of jam is 1890 g.
The total weight of 5 tins of beans is 750 g.
Work out the weight of

   a 1 tin of beans   .... **(2 marks)**   **b** 1 jar of jam.   .... **(2 marks)**

5 Three companies sell the same type of furniture.
The price from Company A in the UK is £1660.
The price from Company B in Germany is €1980.
The price from Company C in the USA is $2250.
The exchange rates are £1 = €1.20 and £1 = $1.35
Which company sells this furniture at the lowest price?
You must show your working.   .... **(3 marks)**

6 Emily, Finn and Grace share some money in the ratio 2 : 3 : 7.
Emily gets £24. How much money does Finn get?   .... **(2 marks)**

7 Ben and his family are going on holiday to France.
The exchange rate is £1 = €1.20.
Ben changes £800 into euros (€).

   a How many euros does Ben get?   .... **(2 marks)**

   In France, a smartphone costs €540.
   In England, the same smartphone costs £460.

   b What is the difference between the cost of the smartphone in
      France and in England in pounds? You must show your working.   .... **(3 marks)**

Now that you have completed this unit, how confident do you feel?

1 Simplifying and using ratio   **2** Using proportion   **3** Inverse proportion

⑦ Averages and range

This unit will help you to calculate and understand averages and range.

AO1 Fluency check

① Write the numbers in numerical order. Start with the smallest.

53 51 42 58 57 55 54

42,51,53,54,55,57,58

② The bar chart shows the favourite crisp flavours of students in Matt's class.

What is the most popular crisp flavour in Matt's class?

Salt and Vinegar

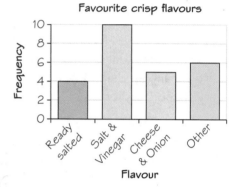

Favourite crisp flavours

③ **Number sense**

Work out

a 6 + 7 + 4 + 9 + 6 + 3

35

b 93 − 38

55

c 35 + 29 + 38 + 32 + 36

170

d 152 − 96

56

e 84 ÷ 6

f 144 ÷ 16

Key points

| An average is a number that represents a set of values. | Mean, median and mode are three types of average. | The range represents how spread out the data is. |

These **skills boosts** will help you to calculate and interpret the mean, median, mode and range.

1 Averages and range   **2** Comparing two distributions   **3** Data in tables

You might have already done some work on averages and range. Before starting the first skills boost, rate your confidence using each concept.

① Work out the mean, median, mode and range of this data set.

14   17   15   11
15   12   16

② Football team A has a median height of 1.81 m and range 15 cm. Football team B has a median height of 1.78 m and range 11 cm. Compare the two football teams.

③ The table shows the number of brothers and sisters for Year 11 students. Work out the mean.

Number of brothers and sisters	Frequency
0	45
1	67
2	26
3	12

How confident are you?

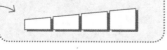

1 Averages and range

To find the mean of a set of data, add up every value and divide by the number of values.
To find the median, write all the values in order and find the value in the middle.
The mode is the most commonly occurring value. Not all data sets have a mode.
To find the range, subtract the smallest value from the biggest value.

Guided practice

Here is a data set.

51   46   52   48   51

Work out

a the mean   **b** the median   **c** the mode   **d** the range.

Worked exam question

---

a Add up all the values.
Total = 51 + 46 + 52 + 48 + 51 = 248
Divide the total by the number of values.

Mean = 248 ÷ 5 = 49.6

$$\text{Mean} = \frac{\text{sum of values}}{\text{number of values}}$$

b Write the values in order. Start with the smallest.

46   48   51   51   52

Write the middle value. This is the median.
Median = 51

c Write the value which appears most often. This is the mode.
It can be easier to identify when the values are in order.

Mode = 51

In this data set, which value appears in the list twice?

d Subtract the smallest value from the biggest value.
The biggest and smallest values can be easier to identify when the values are in order.

Range = 52 − 46 = 6

---

① Here are the numbers of text messages that Sara sends each day for five days.

13   17   14   17   15   13, 14, 15, 17, 17.

Work out

a the mean
15.2

b the median
15

c the mode
17

d the range.
17 − 13 = 4

② Here are the heights, in cm, of a group of 6-year-old children.

125   128   121   131   130   129   132

Work out   121, 125, 128, 129, 130, 131, 132

Hint Not all data sets have a mode.

a the mean
124.14

b the median
129

c the mode
No mode

d the range.
11

(3) Here are the masses, in grams, of some letters.

24   21   27   24   23   22

Hint For an even set of values, the median is the value halfway between the two middle

values: $\frac{23 + 24}{2} = \square$

Work out   21, 22, 23, 24, 24, 27

a the mean

23.5

b the median

23

c the mode

24

d the range.

6

(4) Phil measured the lengths of some bananas in cm.

19.8   19.7   18.5   19.1   18.9   19.2

Work out

a the mean

b the median

c the mode

d the range

(5) Jay went on a hiking holiday to the Yorkshire Dales. The table shows the distances Jay walked each day.

Work out

Day	Mon	Tue	Wed	Thu	Fri
Distance (km)	8.5	11.2	15.3	13.4	7.1

a the mean distance Jay walked each day

b the median distance Jay walked each day

c the range of distances Jay walked each day

(6) Write the mode for the data displayed in each bar chart.

a

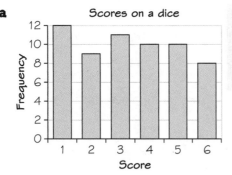

Hint Which score has the highest frequency?

b

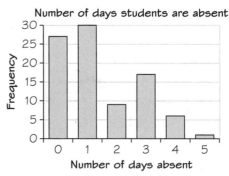

Exam-style question

(7) Here is a data set.   4   7   3   3   6   4

a Find the median.   (2 marks)

b Work out the range.   (2 marks)

c Work out the mean.   (2 marks)

Reflect   Use the definitions of the averages and range on page 44 to write new lyrics to a favourite tune. Learn your new lyrics to help you remember the definitions.

2 Comparing two distributions

Mean, median and mode show the average for the data.
Range shows how much the data is spread out.
To compare two sets of data you need to compare both an average and a measure of spread.

Guided practice

There are 13 girls and 12 boys in a class.
Here are the heights, in cm, of the girls.

159 172 149 155 153 165 178 165 163 175 169 171 166

Here are the heights, in cm, of the boys.

158 160 163 167 169 170 174 174 175 176 180 184

Compare the heights of the boys with the heights of the girls.

_____

Write the heights of the girls in order, starting with the smallest.

Girls:

149 153 155 159 163 165 165 166 169 171 172 175 178

Use the ordered numbers to find the median.

Median height for girls = ...165........ cm

Write the median height for the boys.

Median height for boys = ...173...... cm

The heights of the boys are already in order.

Work out the ranges for the girls and the boys.

Range for girls = ...178.... − ....149....

= 29 cm

Range = tallest height − shortest height

Range for boys = ...184.... − ....158....

= 26 cm

Compare the medians of the data sets. Who, on average, is taller?

On average, the ......Girls...... are

(taller than)/shorter than/the same as the ....boys.....

Select the correct comparison.

Compare the ranges of the data sets. Whose heights are more spread out?

The heights of the ....boys.... are (more)/less/similarly

Select the correct comparison.

spread out (than)/as the heights of the ....girls....

---

① Here are the distances, in metres, that 12 men jumped in the 2016 Olympic long jump final.

8.05 8.29 7.82 8.10 8.38 7.87 8.17 7.86 8.06 8.37 7.97 8.25

a Work out

i the median

7.98

ii the range.

8.38 − 7.82 = 0.56

In the long jump final for women, the median was 6.79 m and the range was 0.59 m.

b Compare the distances that the men jumped with the distances that the women jumped.

....

....

....

....

② Here are the times, in seconds, for the 100 m final in the men's T13 category in the 2016 Paralympics.

11.01   10.83   11.45   10.78   11.05   11.00   10.64   11.38

The times in the men's T44 category were

10.81   11.02   11.03   11.11   11.16   11.17   11.26   11.33

Compare the times of the T13 men with the times of the T44 men.

....

....

....

....

....

....

....

Exam-style question

③ Mrs Ward asked each student in her class to record the number of times they used a mobile phone at the weekend.

Here are the results for the boys.

7   2   8   9   7   9   8   4   3

a Work out the median.

.... **(2 marks)**

Here are the results for the girls.

15   8   9   9   12   14   15

b Compare the numbers of times the boys used a mobile phone with the numbers of times the girls used a mobile phone.

....

....

....

....

.... **(4 marks)**

Reflect   Explain why you need to look at both average and range when comparing two sets of data.

③ Data in tables

Guided practice

The table shows the results of spinning a 5-sided spinner.
Work out the mean result.

Score on spinner	Frequency
1	11
2	9
3	5
4	8
5	7

Add extra columns to the table to show your working.

Work out the score × frequency for each score.

Add an extra row to work out the totals.

Score on spinner	Frequency	What the information means	Score × frequency
1	11	The spinner landed on 1 eleven times	1 × 11 = 11
2	9	The spinner landed on 2 nine times	2 × 9 = 18
3	5	The spinner landed on 3 5	3 × 5 = 15
4	8	The spinner landed on 4 eight times	4 × 8 = 32
5	7	The spinner landed on 5 seven times	5 × 7 = 35
	40		111

Work out the total number of spins.
This is the frequency column.

Work out the total for each number
in the score × frequency column.

Calculate the mean.

Mean = ~~111~~ 40 ÷ ~~40~~ 40

= 2.775

$$\text{Mean} = \frac{\text{total of all the results}}{\text{total number of spins}}$$

① The table shows the results of rolling a 4-sided dice.

Score on dice	Frequency	What the information means	Score × frequency
1	13	The spinner landed on 1 thirteen times	1 × 13 = 13
2	14		2 × 14 = 28
3	16		3 × 16 = 48
4	16		4 × 16 = 64
	59		153

Work out the mean score.

153 ÷ 59 = 2.59

2 The table gives information about the number of people living in some households.

Number of people	Frequency	what the information means	score x Frequency
2	7		2×7=14
3	12		3×12=36
4	19		4×19=76
5	8		5×8=40
6	3		6×3=18
7	1		7×1=7
	50		191

Work out the mean number of people per household.

191 ÷ 50 = 3·82

(Exam-style question)

3 Maisie works in a café.

At noon one day she records the number of customers sitting at each table in the café.
Here are her results.

Number of customers sitting at a table	Frequency		score x Frequency
0	3		0×3=0
1	4		1×4=4
2	12		2×12=24
3	8		3×8=24
4	7		4×7=28
5	1		5×1=5
	35		85

Work out the mean number of customers at a table.

85 ÷ 35 =   2·4    **(2 marks)**

4 The grouped frequency table gives information about the heights of 30 students.

Height, h (cm)	Frequency
$150 < h \leqslant 160$	3
$160 < h \leqslant 170$	11
$170 < h \leqslant 180$	14
$180 < h \leqslant 190$	2

Hint
The modal class interval is the class interval with the highest frequency.

a Write down the modal class interval. $170 < h \leqslant 180$

Hint
The interval containing the median is the class interval that has the middle height in it.
If the 30 students are lined up in order of height, where will the middle be?

b Write down the interval containing the median. $170 < h \leqslant 180$

c Work out an estimate for the range.   40 cm

Hint
What is the smallest possible height?
What is the greatest possible height?

(Reflect) For Q4c, why can you only estimate the range and not work it out exactly?

Practise the methods

Answer this question to check where to start.

Check up

Tick the median for this data set.

~~35~~   ~~34~~   38   ~~34~~   ~~37~~   34, 34, 35, 37, 38

(A) ○	(B) ○	(C) ○	(D) ○	(E) ✓
38	4	34	35.6	35

If you ticked E, find the mean, mode and range. Then go to Q2.

If you ticked A, B, C or D go to Q1 for more practice.

---

(1) Here are the masses, in grams, of some plums.

~~55~~   58   ~~53~~   ~~55~~   ~~56~~   ~~57~~   ~~51~~

a Write the numbers in order, starting with the smallest.

51, 53, 55, 55, 56, 57, 58

b Use your answer for part **a** to write the median.

55

c Work out the mean.

~~358~~ 385 6/8

55

d Write the mode.

55

e Work out the range.

58 − 51

7

Exam-style question

(2) Here are 10 students' estimates of a 13 cm line.

13.2   12.1   13.6   12.8   12.5   13.0   12.8   13.3   12.9   13.5

~~12 cf~~

a Write down the mode.

.... (1 mark)

b Work out the range.

.... (2 marks)

c Work out the median.

.... (2 marks)

d Work out the mean.

.... (2 marks)

3 Here are the masses, in kilograms, of 10 boys in a class.

55	53	56	55	64	55	56	50	53	54

a For these masses, find

i the median ....

ii the range. ....

The masses of the 15 girls in the class are recorded.
The median is 53 kg and the range is 12 kg.

b Compare the masses of the boys with the masses of the girls.

....

....

4 The table shows information about the number of emails a group of students sent during a day.

Number of emails	Frequency
0	7
1	8
2	5
3	5
4	4
5	1

Work out the mean number of emails sent per student.

....

5 The grouped frequency table gives information about the weights of some apples.

Weight, w (g)	Frequency
$100 < w \leqslant 110$	6
$110 < w \leqslant 120$	29
$120 < w \leqslant 130$	14
$130 < w \leqslant 140$	1

a Write down the modal class interval.

....

b Write down the interval containing the median.

....

c Work out an estimate for the range.

....

Problem-solve!

Exam-style questions

1. Sarah recorded the minimum temperature, in °C, on seven days in January.
   Here are her results.

Day	Monday	Tuesday	Wednesday	Thursday	Friday	Saturday	Sunday
Temperature (°C)	2	−2	−1	2	−1	3	4

   a Work out the range of the temperatures. .... (2 marks)

   b Work out the mean of the temperatures Sarah recorded.

   .... (2 marks)

2. Here are four number cards.
   One of the cards is turned over so you cannot see its number.
   The mean of the four numbers is 8.
   Work out the number you cannot see. .... (3 marks)

   $\boxed{9}\ \boxed{\ }\ \boxed{6}\ \boxed{8}$

3. Sameer has three cards. Each card has a number on it. The numbers are hidden.
   The mode is 10. The mean is 11.
   Work out the numbers on the cards. .... (2 marks)

4. Dan, Liz and Oscar buy some sweets.
   Dan buys x sweets.
   Liz buys 5 more sweets than Dan.
   Oscar buys twice as many sweets as Dan.
   Find an expression, in terms of x, for the mean
   number of sweets Dan, Liz and Oscar buy. .... (2 marks)

5. Ryan rolled an ordinary dice 30 times.
   The frequency table shows his results.
   Ryan worked out the mean score as 8.

   a Explain why it is impossible for the mean score to be 8.

Score	Frequency
1	5
2	4
3	7
4	6
5	4
6	4

   ....
   .... (1 mark)

   Eden also worked out the mean score.
   Here is her working.
   $1 \times 5 + 2 \times 4 + 3 \times 7 + 4 \times 6 + 5 \times 4 + 6 \times 4 = 102$
   $102 \div 6 = 17$
   The mean score is 17.

   b Describe the mistake Eden made in her working.

   .... (2 marks)

Now that you have completed this unit, how confident do you feel?

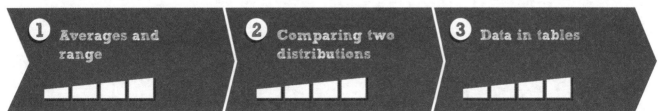

① Averages and range

② Comparing two distributions

③ Data in tables

Data collection

This unit will help you to design and use data collection sheets and understand sampling.

AO1 Fluency check

(1) Write down the number of tallies.

a ‖‖ ‖ ..... *7*   **b** ‖‖ ‖‖ ‖‖ ..... *13*   **c** ‖‖ ‖‖ ‖‖ ‖ *16*

(2) Write three numbers that satisfy each inequality.

a $n > 4$ **b** $n \leq 4$ **c** $0 < n \leq 3$ **d** $-2 \leq n < 4$

5, 6, 7   *2, 3, 4*

(3) Number sense

Work out

a $37 + 28 + 15 =$ *80*   **b** $52 + 14 + 29 =$ *95*   **c** $27 + 62 + 35 =$ *124*

Key points

There are two types of data: discrete data and continuous data.

Discrete data is particular values, for example the number of text messages you receive over a week. You group discrete data in groups like 1–10, 11–20, 21–30, etc.

Continuous data is measured data and can have any value, for example time, length and weight. You group continuous data using inequalities, making sure that there are no gaps or overlaps between the groups.

These **skills boosts** will help you to understand and use class intervals, design and use data collection sheets and understand sampling and bias.

1 Data collection sheets   **2 Bias**

You might have already done some work on data collection. Before starting the first skills boost, rate your confidence using each concept.

(1)
Rashida records the height, to the nearest cm, of 15 students.
156, 172, 165, 181, 168, 175, 169, 170, 171, 163, 170, 168, 169, 171, 166
Design and complete a data collection sheet for Rashida's data.

(2)
A headteacher wants to know students' views on the new school uniform. She plans to give a questionnaire to the boys in Year 7. Will this method give fair or biased results? Explain.

How confident are you?

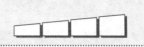   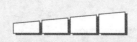

1 Data collection sheets

The four inequality signs are: > meaning 'greater than', < meaning 'less than', ⩾ meaning 'greater than or equal to', and ⩽ meaning 'less than or equal to'.

A data collection sheet is a table used to record data.

Guided practice

Aleska asks students how long it takes them to travel to school, to the nearest minute.

Here are her results.

| 5 | 12 | 15 | 7 | 4 | 7 | 10 | 8 | 9 | 23 |
| 7 | 10 | 5 | 3 | 3 | 8 | 9 | 6 | 12 | 17 |

Design and complete a suitable data collection sheet for Aleska's data.

Draw a table.

Choose the groups so that there are four to six groups.

The measurement of time is continuous data, so use inequalities.

Make a tally mark for each data item in the correct row.

Write the total for each tally in the frequency column.

Time, t (minutes)	Tally	Frequency
$0 < t \leqslant 5$	I	
$5 < t \leqslant$		

Worked exam question

The first student takes 5 minutes to travel to school. This goes into the $0 < t \leqslant 5$ group, not the $5 < t \leqslant 5$ group.

① Patek records the mark that each student in his maths group achieved in a test.
46, 32, 19, 56, 45, 30, 48, 51, 44, 47, 42, 50, 45, 41, 47, 47, 43, 50, 39, 46
Design and complete a suitable data collection sheet for this information.

Hint
This is discrete data. Group discrete data like 1–10, 11–20, etc. You may not need to use all of the rows in the table.

(2) Renee records the heights, in cm, of some tomato plants.

54.2, 53.1, 60.1, 51.9, 49.8, 50.0, 56.3, 55.0, 54.5, 54.3, 53.7, 47.6, 51.4, 56.5, 53.5

Design and complete a data collection sheet for this information.

Hint
Is this discrete or continuous data?

(3) Candice wants to find the time students in her year group spend completing homework.
Design a suitable table for a data collection sheet that she could use to collect this information.

Exam-style question

(4) Dean wants to find the heights of the students in his year group.
Design a suitable table for a data collection sheet that he could use to collect this information.

(3 marks)

Reflect Using your answers to the questions in this skills boost, explain what a data collection sheet is.

2 Bias

A sample represents the population. The population is the whole group that you are interested in.
A sample that is too small can be biased as it does not represent the whole population.
Questions in a questionnaire should not be biased or leading. Biased or leading questions try to influence your answer.

Guided practice

Oliver is doing a survey to find out how many magazines people buy.
Oliver asks his friends at school to fill in his questionnaire.
This may not be a good sample to use.

a Give two reasons why.

There are 900 students at Oliver's school.

b How many students should Oliver include in his sample?

---

a There may be more than two possible answers.

You need to be able to justify your answers.

> Samples may be biased because of age, gender, interests, sample size, location of sample, etc.

Reason 1: This sample may be biased because ....

....

Reason 2: This sample may be biased because ....

....

b 10% of 900 = ... students

> A good sample size is 10% of the population.

1 Amy is doing a survey to find out how much time students spend playing sport.
She is going to ask the first 10 girls on the register for her PE class.
This may not produce a good sample for Amy's survey.

a Give two reasons why.

....

b There are 780 students at Amy's school.
Circle the most appropriate sample size she should choose.

20   50   80   390   780

c Explain your choice in part **b**.

....

2 The manager of a sports centre is planning a new fitness class.
She wants to know if many members will attend the fitness class.

The manager plans to give the questionnaire to the first 15 members who arrive
at the sports centre on Tuesday morning.

a Give two reasons why this may *not* be a suitable sample.

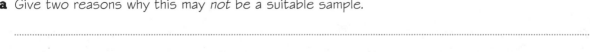

....

Unit 8 Data collection   **55**

b The sports centre has 1540 members.

Circle the most appropriate sample size the manager should choose.

15   50   150   500   1000

c Explain your choice in part **b**.

....

③ Paul wants to find out how much money people spend buying music downloads. He asks 100 people who are listening to music using their headphones in town. Paul's sample is biased. Explain why.

Hint A biased sample does not represent the whole population because not everyone in the population has an equal chance of being chosen.

....

....

Exam-style question

④ Kirsty wants to find out how much time people spend watching rugby on television.

She carries out a survey using a questionnaire.

Kirsty asks the girls in her class to do her questionnaire.

Her sample is biased.

Give *two* reasons why.

....

....   **(2 marks)**

⑤ Helen carries out a survey on healthy eating. She uses these two questions in a questionnaire.

Question 1 What is your age?

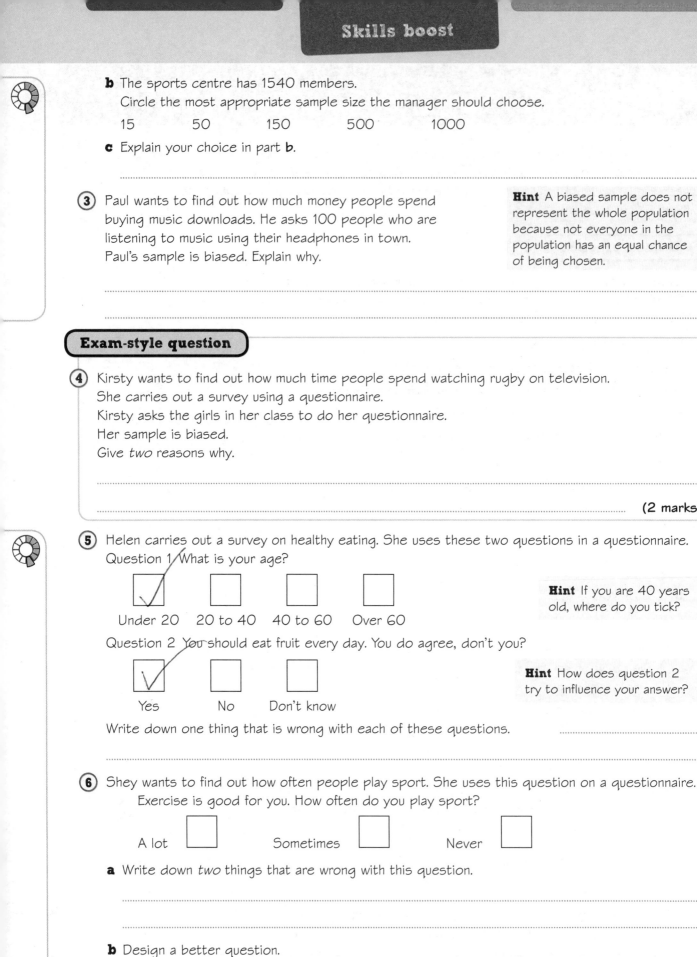

☑   ☐   ☐   ☐

Under 20   20 to 40   40 to 60   Over 60

Hint If you are 40 years old, where do you tick?

Question 2 You should eat fruit every day. You do agree, don't you?

☑   ☐   ☐

Yes   No   Don't know

Hint How does question 2 try to influence your answer?

Write down one thing that is wrong with each of these questions.   ....

....

⑥ Shey wants to find out how often people play sport. She uses this question on a questionnaire.

Exercise is good for you. How often do you play sport?

A lot ☐   Sometimes ☐   Never ☐

a Write down *two* things that are wrong with this question.

....

....

b Design a better question.

....

....

Reflect   Use your answers in this skills boost to list why a sample may produce biased data. Write a list of what makes a good, unbiased question on a questionnaire.

Practise the methods

Answer this question to check where to start.

Check up

Harry measures the heights of students, in metres.

Tick the correct inequalities for his data collection sheet.

A ◯
$1.5 \leqslant h < 1.55$
$1.55 \leqslant h < 1.6$
$1.6 \leqslant h < 1.65, \ldots$

B ◯
$1.5 < h \leqslant 1.55$
$1.55 < h \leqslant 1.6$
$1.6 < h \leqslant 1.65, \ldots$

C ◯
$1.5 < h < 1.55$
$1.55 < h < 1.6$
$1.6 < h < 1.65, \ldots$

D ◯
$1.5 \leqslant h \leqslant 1.55$
$1.55 \leqslant h \leqslant 1.6$
$1.6 \leqslant h \leqslant 1.65, \ldots$

E ◯
$1.5 < h \leqslant 1.55$
$1.56 < h \leqslant 1.60$
$1.61 < h \leqslant 1.65, \ldots$

F ◯
$1.5 > h \geqslant 1.55$
$1.55 > h \geqslant 1.6$
$1.6 > h \geqslant 1.65, \ldots$

If you ticked A or B go to Q2.

If you ticked C, D, E or F go to Q1 for more practice.

1 Ella records the times taken by athletes to run the 100 metres in a competition.
The fastest time is 11.2 seconds, and the slowest time is 13.5 seconds.
List a suitable set of inequalities for Ella's data collection sheet.

....

....

Exam-style question

2 Setter is an internet site. George carries out a survey on the lengths of time people spend using Setter. He uses this question in his questionnaire.

Setter is a very useful internet site. How much time do you spend using Setter?

☐ ☐ ☐ ☐
None   A little   Not much   A lot

a Write down three things that are wrong with this question.

....

....

.... (3 marks)

b Design a better question that George could use.

....

.... (2 marks)

George is going to give his questionnaire to his friends.
c Why is this *not* a good sample?

.... (1 mark)

Problem-solve!

1. Here are three hypotheses.
   - Year 11 students have faster reaction times than Year 7 students.
   - Year 11 students spend more money on magazines than Year 7 students.
   - Year 11 boys in your school spend more time using games consoles than girls in your school do.

   Hint A hypothesis is a statement that you test to see if it is true or not.

   Choose a hypothesis from the list and circle it. For your chosen hypothesis:

   a Write down what data you need to collect to test the hypothesis and where/how you will get the data without your data being biased.

   ....

   ....

   b Explain how you will work out your sample size.

   ....

   c Write a question for the questionnaire to test your chosen hypothesis.

   ....

   ....

   d Draw a data collection sheet for your question in part **c**.

Exam-style questions

2. Ava is doing a survey to find out how often students walk to school. She asks 10 students walking through the school gate.

   a This may not produce a good sample for Ava's survey. Give *two* reasons why.

   ....

   .... (2 marks)

   b Design a suitable question for Ava to use on a questionnaire to find out the number of times students walk to school per week.

   ....

   .... (2 marks)

Now that you have completed this unit, how confident do you feel?

1 Data collection sheets	**2** Bias

⑨ Tables, charts and graphs

This unit will help you to draw dual bar charts and pie charts and to interpret pie charts and misleading graphs.

AO1 Fluency check

① Write down the number of degrees in

  a a full turn ......$360°$......   **b** a right angle ......$90°$......   **c** a straight line ......$180°$......

② Write the shaded amount as a fraction for each shape.

  a [shape] $3/4$ **b** [shape] $1/3$ **c** [shape]

③ **Number sense**

Work out

  a $360 ÷ 4 =$ ....   **b** $360 ÷ 6 =$ ....   **c** $18 × 3 =$ ....   **d** $22 × 5 = 16$

Key points

The height of a bar on a chart represents frequency. Frequency is on the vertical axis.	Data is represented in a pie chart using sectors.

These **skills boosts** will help you to draw dual bar charts and line graphs, to draw and interpret pie charts and interpret misleading graphs.

① **Dual bar charts and line graphs** ② **Interpreting pie charts** ③ **Drawing pie charts** ④ **Misleading graphs**

You might have already done some work on tables, charts and graphs. Before starting the first skills boost, rate your confidence using each concept.

① The table shows information about the subject students like best.

	Boys	Girls
Maths	6	4
English	5	11
Science	13	9
Art	9	9

Draw a suitable diagram or chart for this information.

② The pie chart shows the types of fish Simon caught.
He caught 60 fish in total. How many of the fish were roach.

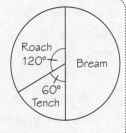

③ The table shows the favourite fruit of 20 students.

Fruit	Frequency
Apple	6
Banana	10
Orange	4

Draw a pie chart for this information.

④ A graph shows a vertical axis of equal steps, labelled 1, 2, 4, 8, 16 and 32. Explain what is wrong with this axis.

How confident are you?

Right margin handwritten numbers: 5, 10, 15, 20, 25, 30, 35, 40, 45, 50, 55, 60, 65, 70, 75, 80, 85, 90, 95, 105, 110, 115

Dual bar charts and line graphs

A dual bar chart compares two sets of data, for example boys with girls. The bars for the same categories are drawn side by side.
A dual bar chart must have a key showing what each bar represents.

Guided practice

Worked exam question

Wanderers and Rovers are two rugby teams.
The table gives information about the total numbers
of points scored by each team over four months.

	January	February	March	April
Wanderers	34	55	42	43
Rovers	47	43	41	56

On the grid, draw a suitable diagram to show this information.

_____

Draw a bar chart with frequency on the vertical axis and months on the horizontal axis.
Choose a sensible scale for the vertical axis.
The highest value is 60, so you could use one large square for every 10 points.
For January, draw the bar for Wanderers up to 34 and the bar for Rovers up to 47.
Draw the bars for the other months.

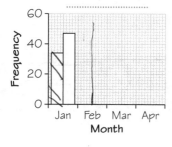

Key

▨ Wanderers

☐ Rovers

Choose how to show each team. You could use two different colours or different patterns (for example, stripes and dots).

Add a key to your chart.

Give your chart a title.

 (1) The table shows information about the average
daily hours of sunshine in London and in Aberdeen
for five months last year.

	May	June	July	August	September
London	5.8	7.3	7.9	5.9	4.5
Aberdeen	4.6	3.9	6.7	5.2	3.4

On the grid, draw a suitable diagram or chart.

Exam-style question

(2) Mark and Megan work for an estate agent.

The table shows information about the numbers of properties each of them sold last year.

	1st quarter	2nd quarter	3rd quarter	4th quarter
Mark	12	27	20	17
Megan	15	21	22	13

The manager wants to compare this information.

On the grid, draw a suitable diagram or chart.

(4 marks)

(3) The line graph shows the average monthly rainfall, in mm, in London.

a Which month has the least rain?

Hint For which month is the line graph the lowest?

March

b Which month has the most rain?

October

c Which two months have the same amount of rain?

April and June

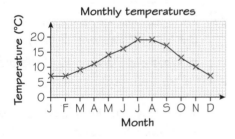

Monthly rainfall

d Between which two months was there the biggest increase in rainfall?

September and october

Hint Between which two months does the line go up the furthest?

e Between which two months was there the biggest decrease in rainfall?

January and February.

(4) The line graph shows the average monthly temperatures, in °C, in London.

a Which months are the coldest?

b Which months are the warmest?

c Between which two months was there the biggest decrease in temperature?

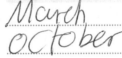

Monthly temperatures

(5) The line graph shows the weight of a baby, in kg, over 10 weeks.

a Between which two weeks did the baby lose weight?

b How much weight did the baby gain in total from being born to 10 weeks old?

c Between which two weeks did the baby weigh 5.5kg?

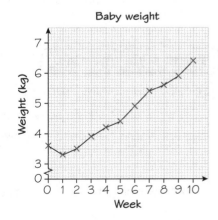

Baby weight

Reflect Shane is planning a visit to London. Using the graphs in Q3 and Q4, which month would you suggest he books his visit for? Explain your answer.

2 Interpreting pie charts

The sectors of a pie chart total 360°.

Guided practice

The pie chart shows some information about the types of food that 80 people like best.

Favourite types of food

a What fraction of the people like English food best?

b Work out the number of people who like Italian food best.

a Measure the angle for the sector labelled English.
Use your protractor.

English = ...°

Write the angle as a fraction of 360°. The total number of degrees for the whole pie chart is 360°.

$$\frac{......................}{360}$$

Simplify your fraction.

$$= \frac{......................}{8}$$

b Measure the angle for the sector labelled Italian.

Italian = ...°

Write the angle as a fraction of 360° and simplify your fraction.

$$\frac{......................}{360} = \frac{......................}{......................}$$

Work out this fraction of the total number of people.

$$\frac{......................}{......................} \text{ of } 80 = 30 \qquad \text{'of' means } \times$$

① The pie chart gives information about what Danny did in 24 hours one Sunday.

a What did Danny spend the most time doing?

....

b Write down the fraction of the pie chart that shows when Danny was working.

....

c How many hours did Danny spend playing football?

....

Saturday activities

Exam-style question

② 120 students are asked if they are given enough homework.
The pie chart shows some information about their answers.

a How many students answered 'enough'?

....

b How many students think they are given too much homework?

....

Homework survey

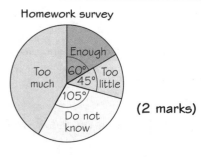

(2 marks)

(3 marks)

Reflect What is the total number of degrees for all the sectors of a pie chart?

3 Drawing pie charts

Guided practice

Miss Jones asks 60 students to name their favourite chocolate.

Here are her results.

Draw an accurate pie chart for her results.

Favourite chocolate	Frequency
Milk	40
White	13
Dark	7

Worked exam question

Draw a new column for the table.

Work out the number of degrees for each type.

The frequencies total 60 and there are 360° in a pie chart.

360 ÷ 60 = 6

Multiply each frequency by 6 to calculate the angle of the sector.

Favourite chocolate	Frequency	Size of angle
Milk	40	40 × 6 = ...°
White	13	13 × .... = ...°
Dark	7	.... × .... = ...°

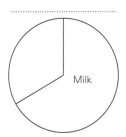

Use your protractor to draw each angle.

Label each sector and give your chart a title.

To draw 240°, you could draw the other two angles and then check that what is left is 240°.

① The table gives information about the results of 45 matches a football team played.

Result	Frequency	Size of angle
Win	17	
Draw	15	
Loss	13	

Draw an accurate pie chart to show this information.

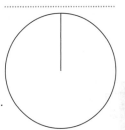

Hint What do you multiply the total number of matches by to get 360?

Hint When drawing the sectors on the pie chart, start from the given line.

Exam-style question

② Sally takes a sample of 60 students at her school.
She asks each student to tell her their favourite type of pizza.
The table shows her results.

Pizza	Number of students
Pepperoni	27
Ham	15
Vegetarian	18

Draw an accurate pie chart to show this information.

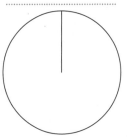

(4 marks)

Reflect The results of a survey show the frequencies 20, 13 and 27. Explain why you cannot use the angles 20°, 13° and 27° to draw a pie chart for this data.

4 Misleading graphs

On any graph, the scales on the axes must go up in equal steps.

Guided practice

The bar chart shows the average house prices in England for July 2015 and July 2016.

A magazine claims that the average house price in England has more than doubled from 2015 to 2016.

Is the claim correct? Explain your answer.

_____

The claim is incorrect because ....

....

....

....

....

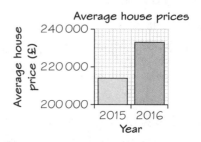

Look at the scale on the vertical axis. Compare the average house price in 2015 with 2016. Has the price doubled? Why does the graph make it look like the price has doubled? The scale starts at £200 000. Where should it have started?

(1) The 3D bar chart shows the total number of weeks four female artists have spent in the top 40 of the UK singles charts.

How could this misleading chart be drawn more clearly?

Hint Can you easily see how many weeks artist 4 spent in the top 40 of the UK singles charts? Can you easily see who spent more time in the top 40, out of artist 1 and artist 4?

....

....

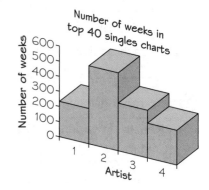

Exam-style question

(2) The graph shows how the price of a loaf of bread changed between 1970 and 1995.

'The graph shows a steady increase in the price of bread.'

Is this statement correct?

Explain your answer.

....

.... (2 marks)

Reflect Explain why the scale on an axis must go up in equal steps. You could use Q2 to help you.

Practise the methods

Answer this question to check where to start.

Check up

Favourite writing implements

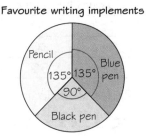

Work out the fraction of students who prefer to use a blue pen.

Tick the correct answer.

 A $\frac{1}{4}$ ○   **B** $\frac{1}{3}$ ○   **C** $\frac{135}{360}$ ○   **D** $\frac{135}{180}$ ○   **E** $\frac{3}{8}$ ○

If you ticked C or E go to Q3.

If you ticked A, B or D go to Q1 for more practice.

(1) Measure the angle of each sector on the pie chart above.

a blue pen = ...°   **b** black pen = ...°   **c** pencil = ...°

(2) Write each angle from Q1 as a fraction of 360°. Simplify your answers.

a blue pen ....   **b** black pen ....   **c** pencil ....

Exam-style questions

(3) The pie chart shows the results of a survey of 100 trees in a wood.

Tree species

a Write the fraction of trees that are beech trees.
Give your answer in its simplest form.

.... (2 marks)

b How many trees are oak trees?

........*108*............ (3 marks)

(4) The table shows the number of medals won by Iran at the 2016 Paralympic Games.

Draw an accurate pie chart to show this information.

Medal	Frequency
Gold	8
Silver	9
Bronze	7

(4 marks)

(5) The dual bar chart shows the median full-time gross earnings per week by gender in the UK.

Hint
'Gross earnings' means earnings before tax is paid.

A report claims that from 2006 to 2010 the weekly salary for men was more than double the weekly salary for women.

Is the report correct? Explain your answer.

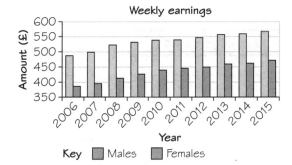
Weekly earnings

Key ■ Males ■ Females

....

....

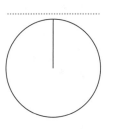

Problem-solve!

1. Jo has used the data in the frequency table to draw a pie chart.

Colour	Frequency
Red	10
Blue	30
Green	50

≠40

Favourite colours

10°

30°

50°

360

Explain what Jo has done wrong.

he has not done 360 of 10 of 30 and 50

2. A bag contains orange, red, yellow and green sweets.

   The pie chart gives information about the numbers of sweets of each colour.

   a What fraction of the total number of sweets in the box are red?

   3⏑10 **(1 mark)**

   In the pie chart, the angle used for yellow is twice the angle used for green.

   b Write down an expression, in terms of x, for the angle used for yellow.

   (1 mark)

   There are 48 sweets in the bag.

   c Work out the number of orange sweets in the bag.

   (2 marks)

   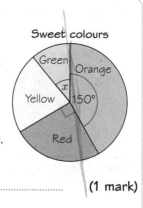

   Sweet colours

   Green, Orange, x, 150°, Yellow, Red

3. Students in Year 11 were asked whether they owned a mobile phone, a tablet or both a mobile phone and a tablet.

   The dual bar chart shows the results for boys and girls.

   All 103 Year 11 boys and 112 Year 11 girls took part in the survey.

   a How many boys own both a mobile phone and a tablet?

   b How many girls own both a mobile phone and a tablet?

   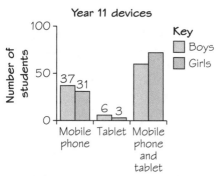

   Year 11 devices

   Key: Boys, Girls

   37 31 6 3

4. Rob and Lionel are tennis players.
   The pie charts show information about the numbers of matches Rob and Lionel each won and lost in tournaments last year.

   Rob played 30 matches. Lionel played 80 matches.

   Lionel won more matches than Rob. How many more?

   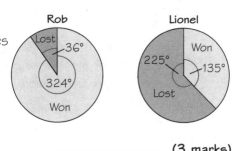

   Rob — Lost, 36°, 324°, Won

   Lionel — Won, 225°, 135°, Lost

   (3 marks)

Exam-style questions

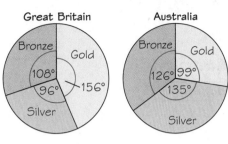
Great Britain   Australia

5) The pie charts show some information about the numbers of medals won by Great Britain and Australia in a sporting event.

Great Britain won 45 bronze medals.

a How many gold medals did Great Britain win?

.... **(2 marks)**

b A reporter says,

'The pie charts show that Australia won more silver medals than Great Britain.'

Is the reporter right? Explain your answer.

....

.... **(1 mark)**

6) The pie chart shows information about the time Michelle spent working in her garden one month.

a What fraction of the time did Michelle spend cutting the grass?

Michelle's gardening jobs

.... **(1 mark)**

Michelle spent 7 hours weeding.
b How much time did Michelle spend planting?

.... **(3 marks)**

This pie chart shows information about the time Dave spent working in his garden in the same month.

Dave says,

'The pie charts show that I spent less time weeding than Michelle did.'

Dave's gardening jobs
Cutting the grass
45°
Planting
80° 180° Weeding
55°
Digging

c Is Dave correct? Explain your answer.

....

.... **(1 mark)**

Now that you have completed this unit, how confident do you feel?

1 **Dual bar charts and line graphs**   ☐☐☐☐

2 **Interpreting pie charts**   ☐☐☐☐

3 **Drawing pie charts**   ☐☐☐☐

4 **Misleading graphs**   ☐☐☐☐

Answers

Unit 1 Multiplying

AO1 Fluency check

① **a** 2 hundreds   **b** 2 tens
   c 2 units   **d** 2 thousands
② **a** 2 units   **b** 2 tenths
   c 2 thousandths   **d** 2 hundredths

③ Number sense

   a 300   **b** 450   **c** 2000   **d** 6700

Confidence check

① 5.83   ② 28.32   ③ −63

Skills boost 1 Multiplying by 10, 100, 1000, 0.1 and 0.01

Guided practice

a $14.52 \times 1000 = \underline{14\,520}$

b

	Th	H	T	U	.	$\frac{1}{10}$	$\frac{1}{100}$
	9	6	1	7			
× 0.1 =		9	6	1	.	7	
× 0.01 =			9	6	.	1	7

$9617 \times 0.01 = 96.17$

① **a** 37   **b** 18   **c** 5640
   d 38.5   **e** 937   **f** 7600
② **a** 2   **b** 45   **c** 600
③ **a** 30   **b** 4.2   **c** 73.5
   d 0.91   **e** 4   **f** 2.96
   g 0.52   **h** 4.721   **i** 0.038
④ 3864 New Taiwan dollars

Skills boost 2 Multiplying decimals by a single-digit number

Guided practice

```
    5  8
 ×     6
 ─────────
 3  4  8
    4
```

$58 \times 6 = \underline{348}$
$5.8 \times 6 = 34.8$

① **a** 34.4   **b** 36   **c** 45.9
   d 44.8   **e** 41.3   **f** 55.2
② **a** 16.35   **b** 33.67   **c** 44.28
   d 46.44   **e** 64.56   **f** 28.36
③ **a** 66.5   **b** 49.44   **c** 13.36
   d 29.2   **e** 18.96   **f** 34.2
④ **a** £29.10   **b** £41.50

Skills boost 3 Multiplying negative integers

Guided practice

$3 \times 4 = \underline{12}$
positive × negative = <u>negative</u>
$3 \times -4 = -12$

① **a**

×	−4	−3	−2	−1	0	1	2	3	4
4	−16	−12	−8	−4	0	4	8	12	16
3	−12	−9	−6	−3	0	3	6	9	12
2	−8	−6	−4	−2	0	2	4	6	8
1	−4	−3	−2	−1	0	1	2	3	4
0	0	0	0	0	0	0	0	0	0
−1	4	3	2	1	0	−1	−2	−3	−4
−2	8	6	4	2	0	−2	−4	−6	−8
−3	12	9	6	3	0	−3	−6	−9	−12
−4	16	12	8	4	0	−4	−8	−12	−16

   b All the numbers in the pink boxes are zero.
   c All the numbers in the blue boxes are negative.
   d All the numbers in the white boxes are positive.
② **a** −15   **b** 22   **c** −36   **d** −42
   e 99   **f** −18   **g** −16   **h** 96
③ **a** −3.7   **b** −6.12   **c** 83.57
   d −19.2   **e** −63.9   **f** −40.5
   g 33.5   **h** −35.6   **i** 5.4
④ **a** −27   **b** 84

Practise the methods

① **a** 30   **b** 18   **c** 126.2
② **a** 500   **b** 382   **c** 210
③ **a** 8000   **b** 12350   **c** 5200
④ **a** 1.7   **b** −0.32   **c** 0.56
⑤ **a** 13.4   **b** 36.9   **c** −3.48
⑥ **a** 5700   **b** 34.3

Problem-solve!

① 130
② £23.67
③ Newcastle
④ No, the parcels will cost £95.67.
⑤ Mia earned £54 and Niall earned £51.75, so Mia earned more.

Unit 2 Dividing

AO1 Fluency check

① **a** 4   **b** 9   **c** 9
   d 8   **e** 7   **f** 6
② **a** 23   **b** 15   **c** 24   **d** 23

③ Number sense

   a 0.5   **b** 4.7   **c** 1.27
   d 35.42   **e** 3.689   **f** 0.23
   g 96.758   **h** 0.0356   **i** 4.0071

Confidence check

① 53.125   ② 1.2575   ③ 710   ④ −2.17

Skills boost 1 Dividing 3-digit numbers by a single-digit number

Guided practice

$$8 \overline{)4\,{}^46\,{}^62\,.\,{}^60\,{}^40} = 0\,5\,7\,.\,7\,5$$

$462 \div 8 = 57.75$

① **a** 62   **b** 158   **c** 73
   d 108   **e** 259   **f** 94

② **a** 157.75   **b** 170.6   **c** 70.25
   d 235.5   **e** 221.75   **f** 45.625

③ £43.25

④ 13.5

Skills boost 2 Dividing decimals by a single-digit number

Guided practice

$$5 \overline{)8\,{}^32\,.\,{}^27\,{}^25} = 1\,6\,.\,5\,5$$

$82.75 \div 5 = 16.55$

① **a** 4.9   **b** 5.3   **c** 2.31
   d 68.49   **e** 3.17   **f** 2.17

② **a** 1.46   **b** 1.55   **c** 2.55
   d 1.1625   **e** 1.8075   **f** 1.693 75

③ £12.37

④ 1.375 m

⑤ £7.85 per hour

Skills boost 3 Dividing by 0.1 and 0.01

Guided practice

	H	T	U	.	$\frac{1}{10}$
			3	.	6
× 10		3	6		
× 100	3	6	0		

$3.6 \div 0.01 = 360$

① **a** 520   **b** 2570   **c** 31
   d 687   **e** 90.2   **f** 321.4

② **a** 400   **b** 3800   **c** 5100
   d 290   **e** 674   **f** 1286.1

③ **a** 1   **b** 90.1   **c** 5200.1
   d 4   **e** 1001   **f** 50

④ 50 portions

⑤ **a** 90   **b** 30

Skills boost 4 Dividing negative integers

Guided practice

a $42 \div 7 = 6$

positive ÷ negative = <u>negative</u>

$42 \div -7 = -6$

b
$$4 \overline{)3\,.\,{}^35\,{}^32} = 0\,.\,8\,8$$

negative ÷ positive = <u>negative</u>

$-3.52 \div 4 = -0.88$

① **a** −10   **b** 9   **c** 6   **d** −8   **e** −8   **f** −8

② **a** −5   **b** −6   **c** −24   **d** 5   **e** −42   **f** 7

③ **a** −42   **b** 90   **c** −1.33
   d −4.18   **e** 1.23   **f** −6.675

④ **a** −3   **b** 5

Practise the methods

① **a** 41   **b** 7   **c** 50   **d** 260

② **a** −83   **b** −25   **c** 540   **d** −910

③ **a** 2.67   **b** 1.92   **c** 0.58   **d** 0.92

④ **a** 2.56   **b** −17.3   **c** −0.57   **d** 8.1

⑤ **a** 2.775   **b** 6.56   **c** −1.4025   **d** 7.0875

⑥ **a** 80   **b** 1.24   **c** −0.625

Problem-solve!

① Wednesday and Thursday

② 174 seats and 29 tables

③ 27 boxes

④ the 6-pint carton

Unit 3 Fractions

A01 Fluency check

① **a** 2   **b** 4   **c** 9   **d** 12

② **a** $\frac{3}{5}$ **b** $\frac{2}{3}$ **c** $\frac{2}{5}$ **d** $\frac{4}{5}$ **e** $\frac{3}{8}$ **f** $\frac{3}{7}$

③ **Number sense**

   a 3, 6, 9, 12, …   **b** 5, 10, 15, 20, …
   c 7, 14, 21, 35, …   **d** 12, 24, 36, 48, …

Confidence check

① $\frac{14}{3}$ ② $1\frac{3}{20}$ ③ $2\frac{1}{7}$ ④ 10

Skills boost 1 Mixed numbers and improper fractions

Guided practice

a $3\frac{1}{4} = \frac{3 \times 4 + 1}{4}$
$= \frac{13}{4}$

b $\frac{7}{3} = 2\frac{1}{3}$

① **a** $\frac{3}{2}$ **b** $\frac{5}{3}$ **c** $\frac{9}{4}$ **d** $\frac{10}{3}$
   e $\frac{8}{5}$ **f** $\frac{14}{5}$ **g** $\frac{13}{3}$ **h** $\frac{15}{4}$

② **a** $2\frac{1}{2}$ **b** $1\frac{1}{3}$ **c** $1\frac{3}{4}$ **d** $1\frac{4}{5}$
   e $2\frac{3}{4}$ **f** $2\frac{2}{3}$ **g** $2\frac{2}{5}$ **h** $1\frac{5}{7}$

③ **a** $\frac{21}{8}$ **b** $4\frac{3}{5}$

Skills boost 2 Adding and subtracting fractions and mixed numbers

Guided practice

a $\frac{7}{9} - \frac{5}{9} = \frac{7-5}{9}$
$= \frac{2}{9}$

b $\frac{7}{8} + \frac{2}{3} = \frac{21}{24} + \frac{16}{24}$
$= \frac{37}{24}$
$= 1\frac{13}{24}$

① **a** $\frac{4}{5}$ **b** $\frac{3}{7}$ **c** $\frac{4}{9}$

② **a** $\frac{11}{14}$ **b** $\frac{13}{15}$ **c** $\frac{7}{10}$

③ **a** $\frac{3}{6} = \frac{1}{2}$ **b** $\frac{11}{40}$ **c** $\frac{11}{10} = 1\frac{1}{10}$

④ **a** $\frac{17}{12} = 1\frac{5}{12}$ **b** $\frac{49}{40} = 1\frac{9}{40}$ **c** $\frac{37}{18} = 2\frac{1}{18}$

⑤ **a** $\frac{6}{10} = \frac{3}{5}$ **b** $\frac{7}{15}$

Skills boost 3 Multiplying a fraction by an integer

> **Guided practice**

$\frac{2}{3} \times 5 = \frac{2 \times 5}{3}$

$= \frac{10}{3}$

$= 3\frac{1}{3}$

① **a** 5 **b** 2 **c** 5

② **a** 12 **b** 8 **c** 4

③ **a** $\frac{7}{4} = 1\frac{3}{4}$ **b** $\frac{12}{5} = 2\frac{2}{5}$ **c** $\frac{12}{7} = 1\frac{5}{7}$

④ **a** $\frac{15}{6} = 2\frac{1}{2}$ **b** $\frac{3}{2} = 1\frac{1}{2}$ **c** $\frac{21}{15} = 1\frac{2}{5}$

⑤ 15

Skills boost 4 Dividing an integer by a fraction

> **Guided practice**

a $3 \div \frac{1}{4} = 12$

b $4 \div \frac{2}{3} = 6$

c $2 \div \frac{3}{5} = 3\frac{1}{3}$

  Alternatively, $2 \div \frac{3}{5} = 2 \times \frac{5}{3} = \frac{10}{3} = 3\frac{1}{3}$

① **a** 6 **b** 8 **c** 10

② **a** 5 **b** 4 **c** 14

③ **a** $4\frac{1}{2}$ **b** $3\frac{3}{4}$ **c** $5\frac{1}{3}$

④ 15

Practise the methods
Check up

$\frac{11}{10} = 1\frac{1}{10}$

① **a** 6 **b** 10 **c** 25

② **a** $\frac{5}{7}$ **b** $\frac{2}{8} = \frac{1}{4}$ **c** $\frac{7}{9}$

③ **a** $\frac{3}{4}$ **b** $\frac{1}{12}$ **c** $\frac{7}{8}$

④ **a** $1\frac{7}{12}$ **b** $1\frac{7}{45}$ **c** $1\frac{7}{15}$

⑤ **a** $2\frac{1}{4}$ **b** $4\frac{1}{6}$ **c** $1\frac{2}{3}$

⑥ **a** 6 **b** 15 **c** 10

⑦ **a** $2\frac{1}{2}$ **b** $6\frac{2}{3}$ **c** $3\frac{3}{5}$

⑧ **a** $\frac{1}{2}$ **b** $\frac{5}{12}$ **c** 8 **d** 20

⑨ **a** $\frac{1}{6}$ **b** $\frac{3}{8}$ **c** $\frac{1}{3}$

  d 1 **e** 2 **f** 4

Problem-solve!
① You cannot subtract fractions with different denominators. James needs to use equivalent fractions to make the denominators the same, then only subtract the numerators.

② 80 students

③ 14 blue counters

④ $\frac{45}{64}$

⑤ **a** Hugh multiplied the numerator and the denominator by 3. He should only multiply the numerator by 3.

  b $\frac{15}{8} = 1\frac{7}{8}$

Unit 4 Fractions, decimals and percentages

> **A01 Fluency check**

① **a** $\frac{1}{2}$ or 0.5 **b** $\frac{3}{4}$ or 0.75 **c** $\frac{3}{5}$ or 0.6

  d $\frac{7}{10}$ or 0.7 **e** $\frac{3}{10}$ or 0.3 **f** $\frac{9}{5}$ or $1\frac{4}{5}$ or 1.8

> **② Number sense**

Fraction	Decimal	Percentage
$\frac{1}{2}$	0.5	50%
$\frac{1}{4}$	0.25	25%
$\frac{3}{4}$	0.75	75%

Confidence check

① $\frac{9}{20}$ **②** 0.65 **③** 77.5%

④ $\frac{6}{100}$, 0.6, 0.606, $\frac{62}{100}$, 66%

Skills boost 1 Converting between decimals and fractions

> **Guided practice**

a **i** $0.3 = \frac{3}{10}$

  ii $0.72 = \frac{72}{100}$

    $= \frac{18}{25}$

b $\frac{8}{25} = \frac{32}{100}$

U	.	$\frac{1}{10}$	$\frac{1}{100}$
0	.	3	2

$\frac{8}{25} = 0.32$

① **a** $\frac{23}{100}$ **b** $\frac{7}{10}$ **c** $\frac{67}{100}$ **d** $\frac{911}{1000}$

② **a** $\frac{1}{2}$ **b** $\frac{4}{5}$ **c** $\frac{16}{25}$ **d** $\frac{2}{25}$

③ **a** 0.1 **b** 0.7 **c** 0.2 **d** 0.6

  e 0.05 **f** 0.45 **g** 0.85 **h** 0.96

④ **a** 0.3 **b** $\frac{2}{25}$

Skills boost 2 Converting a fraction to a decimal

Guided practice

$$16 \overline{)\,1\,1\,.\,0\,^{14}0\,^{12}0\,^{8}0}$$
$$0\,.\,6\quad 8\quad 7\quad 5$$

$\frac{11}{16} = 0.6875$

1. **a** 0.15　　**b** 0.56　　**c** 0.8
   d 0.175　　**e** 0.575　　**f** 0.2375
2. **a** 0.625　　**b** 0.1875　　**c** 0.78125
   d 0.4375　　**e** 0.675　　**f** 0.7125
3. **a** 1.6　　**b** 1.35　　**c** 1.425
   d 1.28　　**e** 2.625　　**f** 2.5125
4. 0.5625

Skills boost 3 Writing one number as a percentage of another

Guided practice

37 out of 40 $= \frac{37}{40}$

$$40 \overline{)\,3\,7\,.\,0\,^{10}0\,^{20}0}$$
$$0\,.\,9\quad 2\quad 5$$

$\frac{37}{40} = 0.925$

$0.925 \times 100 = 92.5\%$

1. **a** 65%　　**b** 72%　　**c** 86%　　**d** 81.25%
2. **a** 95%　　**b** 37.5%　　**c** 64%　　**d** 77.5%
3. 45%　　4. 10%　　5. 15%
6. 25%　　7. 70%

Skills boost 4 Ordering and comparing fractions, decimals and percentages

Guided practice

$\frac{11}{20} = 20 \overline{)\,1\,1\,.\,0\,^{10}0} = 0.55 \times 100 = 55\%$
$$\phantom{\frac{11}{20} = 20\,)}0\,.\,5\quad 5$$

$0.52 \times 100 = 52\%$

$\frac{1}{5} = 20\%$

$0.505 \times 100 = 50.5\%$

$\frac{1}{5}$, **0.505**, 0.52, **53%**, $\frac{11}{20}$

1. 25%, 0.3, $\frac{1}{2}$, 0.75, 100%
2. 0.35, $\frac{2}{5}$, $\frac{3}{4}$, 80%, 0.9
3. 45%, 44%, 0.404, 0.4, $\frac{1}{4}$
4. School B
5. $\frac{7}{20}$, 36%, 0.37, $\frac{3}{8}$

Practise the methods

1. $\frac{3}{5}$, $\frac{60}{100}$ and 0.6

   $\frac{3}{10}$, $\frac{30}{100}$ and 0.3

   $\frac{3}{20}$, $\frac{15}{100}$ and 0.15

   $\frac{9}{10}$, $\frac{90}{100}$ and 0.9

   $\frac{4}{5}$, $\frac{80}{100}$ and 0.8

$\frac{19}{20}$, $\frac{95}{100}$ and 0.95

$\frac{18}{25}$, $\frac{72}{100}$ and 0.72

$\frac{7}{10}$, $\frac{70}{100}$ and 0.70

2. **a** $\frac{4}{25}$　　**b** $\frac{23}{25}$　　**c** $\frac{11}{20}$　　**d** $\frac{21}{50}$
3. **a** 0.375　　**b** 0.3125　　**c** 0.175　　**d** 0.9375
4. **a** 82%　　**b** 88%　　**c** 55%　　**d** 46.875%
5. 43.75%
6. $\frac{7}{20}$, 36%, 0.37, $\frac{3}{8}$

Problem-solve!

1.

Fractions	Decimals
$\frac{1}{3}$	0.33333…
$\frac{2}{3}$	**0.66666…**
$\frac{1}{9}$	0.11111…
$\frac{2}{9}$	0.22222…
$\frac{5}{9}$	**0.55555…**
$\frac{7}{9}$	**0.77777…**
$\frac{4}{9}$	0.44444…

2. 5%, 0.05 or $\frac{1}{20}$　　3. 15.625%
4. 600 g　　5. Briony
6. 32%　　7. 65.5%, $\frac{2}{3}$, 0.67, $\frac{11}{16}$

Unit 5 Probability

AO1 Fluency check

1. $\frac{7}{20}$　　2. 30%

3. Number sense

　a 0.4　　**b** 0.8　　**c** $\frac{3}{4}$　　**d** $\frac{2}{5}$

Confidence check

1. $\frac{3}{7}$　　2. $\frac{2}{9}$　　3. 360
4. black saloon, black estate, black hatchback,
   red saloon, red estate, red hatchback,
   silver saloon, silver estate, silver hatchback

Skills boost 1 The probability scale

Guided practice

a $P(3) = \frac{1}{3}$

b $P(\text{not } 3) = 1 - P(3)$

$$= 1 - \frac{1}{3}$$

$$= \frac{2}{3}$$

①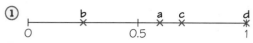

② 0.7

③ $\frac{5}{9}$

④ $\frac{2}{3}$

Skills boost 2 Mutually exclusive outcomes for one event

> **Guided practice**

Number of ways of getting a vowel = 7

Number of possible outcomes = 17

Probability of a vowel = $\frac{7}{17}$

① **a** mutually exclusive

    b not mutually exclusive

    c mutually exclusive

② **a** $\frac{3}{10}$ **b** $\frac{7}{10}$

③ **a** $\frac{7}{20}$ **b** $\frac{13}{10}$

④ **a** $\frac{71}{110}$ **b** $\frac{22}{110}$ or $\frac{1}{5}$

Skills boost 3 Estimating successes

> **Guided practice**

a Estimate for the spinner landing on red

    $= \frac{2}{5} \times 300$

    $= 120$ times

b Estimate for the spinner landing on blue

    $= 0.5 \times 300$

    $= 150$ times

c Estimate for the spinner landing on green

    $= 10\%$ of 300

    $= 30$ times

① 60

② **a** 0.7   **b** 140

③ 60

④ 20

⑤ 20

Skills boost 4 Mutually exclusive outcomes for two events and frequency trees

> **Guided practice**

a

Coin 1	Coin 2
heads	heads
heads	tails
tails	heads
tails	tails

b Number of possible outcomes = 4

    Probability of two tails = $\frac{1}{4}$

① tomato and ham, tomato and cheese, vegetable and ham, vegetable and cheese

② Alfie and Lexi, Alfie and Clara, Alfie and Esme, Tom and Lexi, Tom and Clara, Tom and Esme, George and Lexi, George and Clara, George and Esme

③ **a**

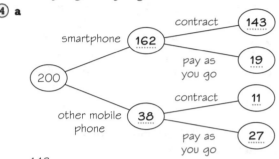

    b   **i** $\frac{3}{9} = \frac{1}{3}$ **ii** $\frac{6}{9} = \frac{2}{3}$

④ **a**

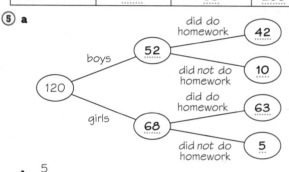

    b $\frac{143}{162}$

    c

	Smartphone	Other mobile phone	Total
Contract	143	11	154
Pay as you go	19	27	46
Total	162	38	200

⑤ **a**

    120 → boys 52 → did do homework 42 / did not do homework 10

    120 → girls 68 → did do homework 63 / did not do homework 5

    b $\frac{5}{68}$

Practise the methods

① **a** 0.7   **b** 0.35   **c** 0.75   **d** 0.17

② 0.55

③ **a** $\frac{5}{12}$ **b** $\frac{7}{12}$

④ **a** 0.2   **b** 64

⑤ 24

⑥ melon and pasta, melon and chicken, melon and fish, soup and pasta, soup and chicken, soup and fish

Problem-solve!

① $\frac{10 - x}{10}$ or $1 - \frac{x}{10}$

② **a** $1 - (0.3 + x) = 0.7 - x$ **b** $300x$

③ **a**

    180 → deep pan 77 → vegetarian 28 / meat 49

    180 → thin crust 103 → vegetarian 36 / meat 67

    b $\frac{49}{77} = \frac{7}{11}$

④ **a** $1 - x$ **b** $100x$

Unit 6 Ratio and proportion

① **a** 1, 2, 5, 10   **b** 1, 2, 3, 4, 6, 12
  c 1, 5, 25   **d** 1, 2, 3, 4, 6, 8, 12, 24
  e 1, 5, 7, 35   **f** 1, 2, 5, 10, 25, 50

② **a** 4   **b** 9   **c** 24

③ **Number sense**

a $\frac{3}{5}$ **b** $\frac{3}{5}$ **c** $\frac{3}{7}$

d $\frac{3}{5}$ **e** $\frac{5}{6}$ **f** $\frac{3}{4}$

Confidence check

① 27 sweets   ② £6.64   ③ $2\frac{1}{2}$ hours

Skills boost 1 Simplifying and using ratio

Guided practice

a 18 : 12 = 3 : 2
b 1 part = 128 ÷ 4 = **32**
  3 parts = **32** × 3
  There are **96** girls in Year 11.

① **a** 4 : 5   **b** 1 : 3   **c** 5 : 4   **d** 4 : 3
  e 3 : 5   **f** 4 : 9   **g** 12 : 5   **h** 7 : 2

② 3 : 1   ③ £39

④ 375 g   ⑤ 104 males

Skills boost 2 Using proportion

Guided practice

1 pen costs £9 ÷ 6 = £**1.50**
10 pens cost £**1.50** × 10
= £15

① £199.76   ② 15.05 m   ③ 5.46 kg
④ £59.75   ⑤ $360   ⑥ 1637.24 lira
⑦ 20 km   ⑧ $\frac{3}{7}$
⑨ **a** 1 : 1.5   **b** $\frac{8}{9}$

Skills boost 3 Inverse proportion

Guided practice

2 builders take 12 hours.
1 builder takes 12 × **2** = **24** hours.
3 builders take **24** ÷ 3 = 8 hours.

① **a** 10 days   **b** 2 days
② 2 days
③ 4 hours
④ 6 minutes
⑤ 6 days
⑥ 6 gardeners
⑦ 3 minutes 20 seconds
⑧ 7.5 hours

Practise the methods

① **a** 4   **b** 2 : 5
② **a** 6   **b** 3 : 5
③ 3 : 4   ④ 85   ⑤ £67.92
⑥ $696   ⑦ 165 cm   ⑧ $\frac{5}{7}$
⑨ 1 : 1.6   ⑩ 3.6 weeks

Problem-solve!

① **a** 3 cm : 7 cm
  b 5 g : 24 g
  c 11 mm : 8 mm
  d 7 ml : 12 ml
② 180 students
③ 2 : 3
④ **a** 150 g   **b** 360 g
⑤ Company A = £1660, Company B = £1650,
  Company C = £1666.67, so Company B is the
  cheapest.
⑥ £36
⑦ **a** €960
  b France = £450, £460 − £450 = £10
  or England = 552 Euros, 552 − 540 = 12 Euros

Unit 7 Averages and range

① 42, 51, 53, 54, 55, 57, 58
② Salt and vinegar

③ **Number sense**

a 35   **b** 55   **c** 170
d 56   **e** 14   **f** 9

Confidence check

① Mean = 14.3, median = 15, mode = 15, range = 6
② On average, players from football team A are 3 cm
  taller than players from team B.
  The heights of the players from team A are more
  spread out than those from team B.
③ 1.03

Skills boost 1 Averages and range

Guided practice

a Total = 51 + 46 + 52 + 48 + 51 = **248**
  Mean = **248** ÷ **5**
    = 49.6
b 46   48   **51**   **51**   **52**
  Median = 51
c Mode = **51**
d Range = **52** − **46**
    = 6

① **a** 15.2   **b** 15   **c** 17   **d** 4
② **a** 128 cm   **b** 129 cm   **c** no mode   **d** 11 cm
③ **a** 23.5 g   **b** 23.5 g   **c** 24 g   **d** 6 g
④ **a** 19.2 cm   **b** 19.15 cm   **c** no mode   **d** 1.3 cm
⑤ **a** 11.1 km   **b** 11.2 km   **c** 8.2 km
⑥ **a** 1   **b** 1 day
⑦ **a** 4   **b** 4   **c** 4.5

Skills boost 2 Comparing two distributions

Guided practice

Girls:
149   153   155   159   **163**
165   **165**   166   169   171
172   175   178
Median height for girls = **165** cm

Median height for boys = 172 cm
Range for girls = 178 − 149
              = 29 cm
Range for boys = 184 − 158
              = 26 cm
On average, the boys are taller than the girls.
The heights of the girls are more spread out than the heights of the boys.

① **a i** 8.08 m
      ii 0.56 m
   b On average, the men jumped a greater distance than the women.
       The distances jumped by the women are a little more spread out than the distances jumped by the men.
② T13 median = 11.005 seconds
   T13 range = 0.81 seconds
   T44 median = 11.135 seconds
   T44 range = 0.52 seconds
   On average, the T13 men are faster than the T44 men running 100 m.
   The times the T13 men can run 100 m are more spread out than the times for the T44 men.
③ **a** 7
   b Girls' median = 12, boys' range = 7, girls' range = 7
       On average, the girls used a mobile phone more times than the boys did.
       The numbers of times the boys and girls used a mobile phone are equally spread out.

Skills boost 3 Data in tables

Guided practice

Score on spinner	Frequency	What the information means	Score × frequency
1	11	The spinner landed on 1 eleven times	1 × 11 = 11
2	9	The spinner landed on 2 nine times	2 × 9 = 18
3	5	The spinner landed on 3 five times	3 × 5 = 15
4	8	The spinner landed on 4 eight times	4 × 8 = 32
5	7	The spinner landed on 5 seven times	5 × 7 = 35
	40		111

Mean = 111 ÷ 40
     = 2.775

① 2.59
② 3.82
③ 2.4 customers
④ **a** $170 < h \leq 180$
   b $170 < h \leq 180$
   c 40 cm

Practise the methods
Check up
Mean = 35.6, mode = 34, range = 4
① **a** 51, 53, 55, 55, 56, 57, 58
   b 55 g   **c** 55 g   **d** 55 g   **e** 7 g
② **a** 12.8 cm **b** 1.5 cm   **c** 12.95 cm **d** 12.97 cm
③ **a i** 55 kg   **ii** 14 kg
   b On average, the boys are heavier than the girls.
       The weights of the boys are more spread out than the weights of the girls.
④ 1.8 emails
⑤ **a** $110 < w \leq 120$
   b $110 < w \leq 120$
   c 40 g

Problem-solve!
① **a** 6°C   **b** 1°C
② 9
③ 10, 10 and 13
④ $\dfrac{x + x + 5 + 2x}{3} = \dfrac{4x + 5}{3}$
⑤ **a** The mean must be less than 6.
   b Eden should have divided by 30, not 6.

Unit 8 Data collection

A01 Fluency check
① **a** 7   **b** 13   **c** 16
② **a** Any three numbers greater than 4, e.g. 5, 6, 7, …
   b Any three numbers less than or equal to 4, e.g. 4, 3, 2, …
   c Any three numbers greater than 0 but less than or equal to 3, e.g. 1, 2, 2.5
   d Any three numbers greater than or equal to −2 but less than 4, e.g. −1, 0, 1, 2, 3

③ Number sense
   a 80   **b** 95   **c** 124

Confidence check
① An example:

Height, h (cm)	Tally	Frequency
$155 < h \leq 160$	|	1
$160 < h \leq 165$	||	2
$165 < h \leq 170$	JHT ||	7
$170 < h \leq 175$	||||	4
$175 < h \leq 180$		0
$180 < h \leq 185$	|	1

② This will give biased results because she is asking all boys from the same age group.

Skills boost 1 Data collection sheets

Guided practice

Time, t (minutes)	Tally	Frequency
$0 < t \leq 5$	ⅠⅠⅠⅠ	5
$5 < t \leq 10$	ⅠⅠⅠⅠ ⅠⅠⅠⅠ	10
$10 < t \leq 15$	ⅠⅠⅠ	3
$15 < t \leq 20$	Ⅰ	1
$20 < t \leq 25$	Ⅰ	1

① An example:

Mark	Tally	Frequency
11–20	Ⅰ	1
21–30	Ⅰ	1
31–40	ⅠⅠ	2
41–50	ⅠⅠⅠⅠ ⅠⅠⅠⅠ ⅠⅠⅠⅠ	14
51–60	ⅠⅠ	2

② An example:

Height, h (cm)	Tally	Frequency
$45 < h \leq 50$	ⅠⅠⅠ	3
$50 < h \leq 55$	ⅠⅠⅠⅠ ⅠⅠⅠⅠ	9
$55 < h \leq 60$	ⅠⅠ	2
$60 < h \leq 65$	Ⅰ	1

③ An example:

Time, t (minutes)	Tally	Frequency
$0 \leq t < 20$		
$20 \leq t < 40$		
$40 \leq t < 60$		
$60 \leq t < 80$		
$80 \leq t < 100$		
$100 \leq t < 120$		

④ An example:

Height, h (cm)	Tally	Frequency
$140 < h \leq 150$		
$150 < h \leq 160$		
$160 < h \leq 170$		
$170 < h \leq 180$		
$180 < h \leq 190$		

Skills boost 2 Bias

Guided practice

a Reason 1: This sample may be biased because **he is asking all of the same age group.**

Reason 2: This sample may be biased because **all of his friends may be the same gender.**

You may also have the reasons that all of his friends have similar interests or that the sample may be too small.

b 10% of 900 = **90** students

① **a** The sample may be biased because it is a small sample size, she is only asking one gender and she is only asking one age group.

b 80   **c** It is close to 10%.

② **a** The sample may be biased because it is a small sample size, they are all arriving on the same day at the same time and so may have similar interests as they may all be turning up for the same event/activity.

b 150   **c** It is close to 10%.

③ He is only asking people who listen to music so are more likely to download music.

④ She is only asking one gender, one age group and in one location.

⑤ Question 1: The response boxes overlap at age 40. Question 2: This is a biased/leading question, as it tells you that you should eat fruit every day, so it may influence your answer.

⑥ **a** This is a biased/leading question, as it tells you that exercise is good for you, so it may influence your answer. The times aren't specified – how often is 'a lot' or 'sometimes'? There is no time scale – is it how often per week/month/etc?

b An example:
How many times a week do you play sport?

None ☐   1-2 times ☐   3-4 times ☐

More than 4 times ☐

Practise the methods

① An example: $11.0 < t \leq 11.5$, $11.5 < t \leq 12.0$, $12.0 < t \leq 12.5$, $12.5 < t \leq 13.0$, $13.0 < t \leq 13.5$

② **a** This is a biased/leading question, as it tells you that Setter is a very useful internet site, so it may influence your answer. The times aren't specified – how long is 'a little', 'sometimes' or 'a lot'? There is no time scale – is it how much time per day/week/month/etc?

b How much time do you spend using Setter?

Zero-10 minutes per day ☐

11-20 minutes per day ☐

21-30 minutes per day ☐

More than 30 minutes per day ☐

c The sample may be biased because it is a small sample size, they are all friends so may have similar interests and they may all be the same age and the same gender.

Problem-solve!

① For the hypothesis 'Year 11 students have faster reaction times than Year 7 students':

a You will need to do an experiment testing and recording the reaction times for Y11 and Y7 students. You will need to take a random sample of the students. Record which year group each student is from and their reaction time.

b A random sample of 10% of each year group would normally be considered sufficient.

c What was the students' reaction time?

d

Reaction time (seconds) Y7 students	Tally	Frequency
$0.10 < t \leq 0.15$		
$0.15 < t \leq 0.20$		
$0.20 < t \leq 0.25$		
$0.25 < t \leq 0.30$		

(2) **a** The sample may be biased because it is a small sample size and she is only asking the students who have walked (since they are walking through the school gate).

b An example: How many times a week do you walk to school?

None ☐   1-2 times ☐   3-4 times ☐

5 times ☐

Unit 9 Tables, charts and graphs

(1) **a** 360°   **b** 90°   **c** 180°

(2) **a** $\frac{3}{4}$ **b** $\frac{1}{3}$ **c** $\frac{7}{8}$

(3) **Number sense**

   a 90   **b** 60   **c** 54   **d** 110

Confidence check

(1)

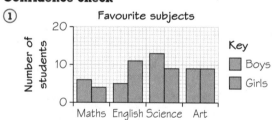

Favourite subjects

(2) 20 fish

(3)

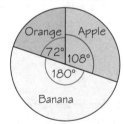

Favourite fruits

(4) The scale must go up in equal steps. 1, 2, 4, 8, 16, 32 is not equal steps; it is doubling.

Skills boost 1 Dual bar charts and line graphs

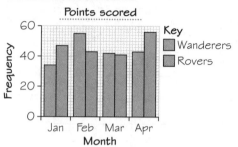

Points scored

(1)

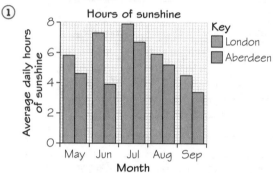

Hours of sunshine

(2)

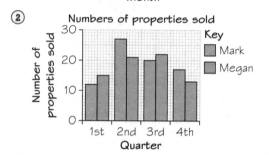

Numbers of properties sold

(3) **a** March
   b October
   c April and June
   d September and October
   e January and February

(4) **a** January, February and December
   b July and August
   c September and October

(5) **a** 0 and 1 week
   b 2.8 kg
   c 7 and 8 weeks

Skills boost 2 Interpreting pie charts

a English = 45°

$$\frac{45}{360} = \frac{1}{8}$$

b Italian = 135°

$$\frac{135}{360} = \frac{3}{8}$$

$$\frac{3}{8} \text{ of } 80 = 30$$

(1) **a** sleeping   **b** $\frac{1}{4}$ **c** 2 hours

(2) **a** 20 students   **b** 50 students

Skills boost 3 Drawing pie charts

> **Guided practice**

$360 \div 60 = 6$

Favourite chocolate	Frequency	Size of angle
Milk	40	$40 \times 6 = \underline{240°}$
White	13	$13 \times \underline{6°} = \underline{78°}$
Dark	7	$\underline{7} \times \underline{6°} = \underline{42°}$

Favourite types of chocolate

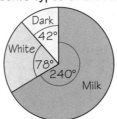

① Football results

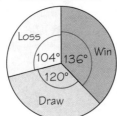

② Pizza survey

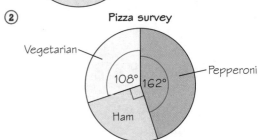

Skills boost 4 Misleading graphs

> **Guided practice**

The claim is incorrect because in 2015 the average house price was £214 000 and in 2016 it was £233 000. Double £214 000 is £428 000, not £233 000. The bar for 2016 looks more than double the bar for 2015 because the scale starts at £200 000 and not 0.

① You cannot read the values for artists 2, 3 and 4 on the scale, and it is difficult to compare the bars because the 3D bar chart is at an angle and on a plain background. So the chart could be drawn more clearly as a 2D bar chart.

② The statement is incorrect. The scale on the 'price of a loaf' axis does not go up in equal steps, but doubles each time. The years do not go up in equal steps, as there is a jump of 10 years from 1970 to 1980 but then the rest are in steps of 5 years.

Practise the methods

① a 135° b 90° c 135°

② a $\frac{3}{8}$ b $\frac{1}{4}$ c $\frac{3}{8}$

③ a $\frac{3}{20}$

  b 30 trees

④ Number of medals

⑤ The report is incorrect. In 2006, the weekly gross earnings were £487 for men and £386 for women. £487 is less than double £386. The bar for men in 2006 looks more than double the bar for women because the scale starts at £350 and not 0.

Problem-solve!

① Jo has used the frequencies as the size of the angles. The frequency totals 90 but a pie chart totals 360°. As $90 \times 4 = 360$, Jo should have multiplied each frequency by 4 to work out the size of each angle.

② a $\frac{1}{4}$ b $2x°$ c 20 sweets

③ a 60 boys b 78 girls

④ 3 matches

⑤ a 65 medals

  b We don't know how many medals Australia won, because the pie chart only shows the proportions of medals won, not how many.

⑥ a $\frac{1}{12}$

  b $1\frac{1}{2}$ hours

  c We don't know because we don't know how long Dave spent in his garden. The pie chart only shows the proportions of time he spent, not the actual amount of time.

1.50

3.6 → $3.60

x (stays same)
1.05 — Flips
       put 0 on end of it
       zero

2.3 ✓ stays same
    ⓪ Flips
       put zero on end.

6.10 ✓ stays same
    ⓪ Flip
       put zero on end.

9.10 ✓ stays same   10.01 $\frac{98}{8}$

to
10.01 ✓ stays same

$$\frac{3}{20} \qquad 0.25 \qquad 3.\grave{}$$

$$25.\zeta$$

|||||
4
8
12
16
20
24
28

$$\overline{IIII}$$
$$\cancel{O} \; O \; O$$
$$O$$

X	100	70	3
4	400	280	R

$$400+$$
$$280+$$
$$12$$
$$\overline{692}$$

20